DISASTER MANAGEMENT IN TELECOMMUNICATIONS, BROADCASTING AND COMPUTER SYSTEMS

DISASTER MANAGEMENT IN TELECOMMUNICATIONS, BROADCASTING AND COMPUTER SYSTEMS

Galal El Mahdy

Mahdi Consult Inc., Canada
Formerly International Telecommunications Union, Switzerland

JOHN WILEY & SONS, LTD

Chichester • New York • Weinheim • Brisbane • Singapore • Toronto

Other Wiley Editorial Offices

John Wiley & Sons, Inc., 605 Third Avenue,
New York, NY 10158-0012, USA

Wiley-VCH Verlag GmbH
Pappelallee 3, D-69469 Weinheim, Germany

John Wiley & Sons Australia, Ltd, 33 Park Road, Milton,
Queensland 4064, Australia

John Wiley & Sons (Asia) Pte Ltd, 2 Clementi Loop #02-01,
Jin Xing Distripark, Singapore 129809

John Wiley & Sons (Canada) Ltd, 22 Worcester Road,
Rexdale, Ontario M9W 1L1, Canada

Library of Congress Cataloguing-in-Publication Data

El Mahdy, Galal.
     Disaster management in telecommunications, broadcasting, and computer systems /
     Galal El Mahdy.
          p. cm.
     Includes bibliographical references and index.
     ISBN 0-471-60812-2 (alk. paper)
     1. Telecommunication systems — Planning. 2. Emergency management. 3.
     Telecommunication systems — Security measures. 4. Computer security. I. Title.

TK5102.5 E375 2001
658.4'77 — dc21

                                                                                    00-043826

British Library Cataloguing in Publication Data

A catalogue record for this book is available from the British Library

ISBN 0-471-60812-2

Typeset in 10/12pt Palatino by Laser Words, Madras, India.
Printed and bound in Great Britain by Bookcraft Ltd, Midsomer Norton, UK.

This book is printed on acid-free paper responsibly manufactured from sustainable forestry in which at least two trees are planted for each one used for paper production.

Murphy's Law

1. Left to themselves, things go from bad to worse

2. Anything that can go wrong, will go wrong

3. If there is a possibility of several things going wrong
the one that will go wrong is the one that will
cause the most damage

4. If everything is going well, watch out!
You have obviously overlooked something

Contents

To Meshmesh

Preface

At graduation I was a young ambitious engineer full of hope, and enthusiasm, willing to change the telecommunications world because — as we all thought at that time — I knew everything that can possibly be taught about telecommunications! As I grew older and gained more hands-on experience, I realized the fallacy of my earlier belief.

I was fortunate to begin my work with a TV and broadcasting organization, where I actively participated in the installation, testing, acceptance, operation, and maintenance of several high-power radio and TV transmitting stations. The catch was that the transmitters were broadcasting political programs, so it was unthinkable from the management's viewpoint that a transmitter should go off the air for any reason. In fact, if the program interruption exceeded five minutes, the shift engineer would be subject to high-level investigation. At an early age, therefore, I was introduced to the wonderful art of disaster management.

I faced different problems when I became director of a research center, responsible for the development and prototype production of some test and measuring equipment for the TV and broadcasting organization, and some academic institutions. This time any delay in the delivery of the outputs would mean delay in implementing projects at both facilities, which meant heavy penalties and the loss of large sums of money.

Changing my career to become manager of a TV picture tube factory, and later on the director general of technical and commercial affairs of a large consumer electronics company, introduced me to different sorts of industrial crises. The TV picture tube factory was going bankrupt due to financial, technical, and administrative problems. I was supposed to effect a turnaround and regain production and lost markets. That task needed quick solutions, and within almost a year the factory was going in the proper direction. Industrial crises occurred for several reasons but the end results were inferior products, improper costing and pricing, and a delay in the production line or the delivery schedule, which meant loss of market and money.

When I became director general of a telecommunications organization, responsible for transforming the country's telecommunications facilities from analog to digital, I faced a lot of telecoms-related disasters. The political influence had an important impact, since how would anyone justify the breakdown of the telecommunications facilities? However, the financial loss due to the disaster was also enormous.

This is why, when I was selected by the United Nations to be the project manager and senior telecommunications training expert for a US$23 million project in the Gulf, I used my courses to stress the topic of telecommunications disaster management.

Also, when I became senior lecturer in telecommunications with the University of Zimbabwe, I prepared and conducted several continuing engineering training programs, in association with the Zimbabwe Institution of Engineers (ZIE), to teach managers, professional engineers, and young graduate engineers the basic tools for detecting, handling, and solving some engineering disasters.

In writing this book, I have relied on these courses that I presented in the Gulf countries, Zimbabwe, and Mozambique, plus the interesting feedback I received from the participants. The book documents the experience gained through working for more than forty years in the field, where I faced a lot of disasters that I had to detect, identify, analyze, and solve. Most of the time I was successful, thank God, but sometimes I was not so fortunate. I hope that this book will help the reader to profit from my experience, and avoid those places where I stumbled.

Dr Galal El Mahdy
January 2001

Acknowledgements

Writing a book is an interesting experience; you actually meet, on paper, the authors and publishers of the books you use, and in time a friendly relationship develops. I have gained new knowledge from all the books listed in the bibliography, yet naturally some of them were closer to me than others. I am indebted to all those authors and publishers for the ideas I gained out of their books. Writing about financial matters was a problem; it meant reading several books about the subject. Perhaps the most useful was Gibson's *Financial Statement Analysis*, published by Thompson Learning; I am grateful for permission to reproduce one of its tables on financial statements.

I would like to thank the International Telecommunications Union (ITU) for granting me permission to use material, and to reproduce a figure from their excellent Training Development Guidelines (TDG). Thanks also go to Fujitsu Limited, Japan, for granting me permission to reproduce a diagram of their digital telephone exchange FETEX-150.

The training courses that I conducted in Zimbabwe, Mozambique, Saudi Arabia, Guinea, Egypt, Qatar, and other Middle Eastern and African countries, were a great help to me in compiling the book. My most recent material is for courses given in Zimbabwe and Mozambique; this was realized with the help of the Zimbabwe Institution of Engineers (ZIE), to whom I am really indebted for their effort, help, and encouragement.

The video training courses listed in the references section are an excellent example of the benefits gained from using new technology in spreading knowledge and expertise. My sincere thanks go to the different producers, presenters, and directors of these videos. I have enjoyed viewing these training videos at the British Council Library in Harare, and in Cairo. These videos gave me a wealth of new ideas and inspiration, which was instrumental in compiling the book. My sincere thanks go to Mr Joseph in Zimbabwe and Ms Cathy Costain in Cairo. I would like also to thank the British Council for a job well done.

Many of my dear friends were kind enough to offer me their help and support. I would like to thank them all for their encouragement, and excellent feedback. Special mentions go to H. E. Fouad Abdel Wahab, Prof. Dr A. M. Khafagi, Prof. Dr A. M. Abdel Latif, Dr Samy Tabbane, Eng. Sherif El Barquoqui, Ms Barbara Wilson, and Mr Hamed Chaabouni.

I would also like to thank the excellent staff at John Wiley & Sons. Special thanks go to Ms Laura Kempster and Ms Sarah Corney for their patience, help, and support.

Finally, I wish to thank my family for understanding and appreciating the reasons behind several missed vacations and lost weekends. Credit also goes to my young grandchildren; without their earnest desire to help, and their innovative computer actions, this book could have been completed much earlier.

Purpose and Scope of the Book

Introduction

Advances in telecommunications technology continue to accelerate. To be competitive in such a dynamic environment, today's managers, professional engineers, and graduate engineers can expect to be challenged daily to keep pace with the technical and operational issues, opportunities and threats surrounding the operation and management of any telecommunications system. This book is written for people who need to be able to detect, understand, handle, and control a telecommunications system during a crisis. In managing disaster, engineers will depend on accurate, accessible, and relevant information to help them think clearly, act strategically, plan effectively, and make decisions that can save their own interests and the interests of their organization.

Major themes of the text are to prepare managers, professional engineers, and graduate engineers to meet the challenges they face in their endeavor to safeguard against disaster, to insure the smooth and safe operation of their telecommunications systems. Managers, and engineers working together in a telecommunications organization are responsible for the success or failure of the organization when a crisis occurs, so the text balances the technical issues with the managerial issues. This theme is punctuated in four ways:

- An integrated use of real-world examples throughout the text.

- Use of feature material from leading telecommunications companies and common carriers, highlighting the issues surrounding disaster and crisis management in a broad spectrum of telecommunications organizations.

- Case studies illustrating how actual disasters were detected, studied, confronted, and successfully controlled (or otherwise).

- Lessons learned from the different examples and case studies.

TWELVE CHAPTERS: TWELVE FLEXIBLE MODULES

Chapter 1

Chapter 1 defines problem, crisis, and disaster. The fundamental idea of thinking about the unthinkable (disaster) is introduced, emphasizing that disaster prevention is not the top management responsibility alone, but it is a responsibility shared by all involved in the organization. The cost of a disaster is discussed at length.

Chapter 2

Since the book is intended for the use of managers, professional engineers, and graduate engineers, it was thought appropriate to highlight some of the fundamental concepts in telecommunications, radio and TV broadcasting, power distribution systems, and information technology, to be used by those readers who need some technical background information about one of these topics. Thus, Chapter 2 provides an overview of the fundamental building blocks in telecommunications systems, including the computer hardware and software, and the TV and radio broadcasting networks. It also gives an insight into the future of telecommunications systems, highlighting the network of the future and the IMT 2000 mobile telecommunications system.

Chapter 3

Chapter 3 deals with common telecommunications systems disasters. It defines minor and major breakdowns and disasters, and it discusses the role played by interruptions, breakdowns, and disasters in the life of a telecommunications organization, stressing their operational and financial impact, on the health of the organization. It goes into greater detail on several topics, including these:

- Basic faultfinding techniques
- Troubleshooting electronic components and circuits
- Cable cuts; theft, fire, and arson
- The computer system, computer crime, the Y2K problem
- Natural disasters such as lightning, earthquakes, tornadoes, floods

Chapter 4

Chapter 4 highlights the fundamentals of disaster management, then introduces the concept of contingency planning along with examples to illustrate its application in the different telecommunications disciplines. Here are some of the topics:

- Analyzing the disaster
- Analyzing the organization
- Contingency planning for different systems
- Proposals for integrated approach plans

Chapter 5

Chapter 5 builds on thinking the unthinkable, a concept from Chapter 1, to stress how disaster management begins with writing the specification. It emphasizes the role of the design engineer in developing specifications, taking into consideration the probability of a disaster. Here are some of the topics:

- Choice of technology
- Facilities and buildings design concepts
- Equipment reliability considerations
- Network configuration and management techniques
- Motivating people, dealing with conflict

Chapter 6

Chapter 6 stresses some of the points raised in the earlier chapters to emphasize their importance. It is impossible to give a disaster recovery plan for each specific telecommunications specialty (around a hundred), so Chapter 6 presents a generic disaster management plan, and each telecommunications manager, engineer, or graduate can adapt it as required. Here are the topics it covers:

- The importance of a disaster planning and recovery strategy
- The disaster prevention scenario
- Risk management
- Vulnerability search, analysis, and rectification
- Contingency management
- The disaster-handling scenario
- A generic crisis management proposal
- Application of the generic disaster management proposal in selected situations

Chapter 7

Chapter 7 complements Chapter 6. Here are the topics it covers:

- Basic reasons for the occurrence of a telecommunications disaster
- Factors affecting the telecommunications disaster
- The requirements of handling a telecommunications disaster
- Disaster management responsibilities
- Cost of a telecommunications disaster
- Security and political issues involved in a telecommunications disaster

Chapter 8

Building on Chapter 5, this chapter looks at how to assess vulnerability in a telecommunications system. It deals with the following topics:

- General considerations
- Assessing technical vulnerability
- Assessing financial vulnerability
- Assessing administrative vulnerability
- Specific action plans for a variety of networks

Chapter 9

Safety is the first and most important consideration when managing a crisis. Chapter 9, deals with the following safety topics:

- The case for a safe system
- Increased profits through engineering safety management
- Safety management as a part of the system
- Firefighting, firecontrol, and fireprevention
- Management of dangerous materials
- Controlling electrical hazards
- Protection from lightning strikes
- Personnel safety considerations
- Environmental hazards

Chapter 10

Legal issues are often neglected in disaster recovery planning. This can have very expensive consequences, since when a disaster happens and is not handled properly, the service is disrupted and customers who have suffered losses might attempt to recover their losses from the telecommunications organization. Chapter 10 looks at liability and limitations to liability, insurance and indemnification.

Chapter 11

Chapter 11 presents real-world situations, I have experienced personally or have been involved with. They cover examples from telephone, data networks, computer networks, TV, and broadcasting networks, and power distribution networks. Each case study shows how actual disasters were detected, studied, confronted, and successfully controlled. The chapter introduces how to deal with case studies, giving the detailed procedure to analyzing them; this is followed by a worked example. Drawn from around the world, the case studies are intended to make the reader think about the case, to understand the whys and hows. Do this before reading the analysis.

Chapter 12

Chapter 12 wraps things up with objectives and goals of the disaster management and recovery plan, the planning process, implementation of the disaster management and recovery plan, case studies and lessons learned.

1

Introduction

OBJECTIVES

- Answer questions at the end of the chapter.
- Categorize and discuss the problems you face at work.
- Enumerate simple problems that may become disasters.
- Describe the shared responsibility of a disaster.
- Give examples of the cost of a disaster at work.

1.1   PROBLEM, CRISIS, DISASTER

1.1.1   Problem

A problem is the cause of an undesired situation; it means that we cannot get the anticipated results from our activities at work. The occurrence of a problem means that incorrect processes and/or procedures are utilized and that the correct ones should be identified and used, since leaving a problem unsolved is the prelude to a disaster.

A problem needs organized and systematic thinking to analyze its causes and decide on the best way to confront it. This is an art in which management qualities, skill and incisive judgment are decisive factors in solving or escalating the problem. Yet a problem is not necessarily negative since it may act as an important element in innovation and development.

Facing problems skillfully and courageously develops the organization and its people's capabilities, creating better managers.

1.1.2 Crisis

A problem that goes on without a solution may be the cause of a crisis. A crisis means a sudden change for the worse (or the better). If a crisis is left unsolved, it develops into disaster. The essence of a true crisis is that the conflict within it rises to a level where there is a real threat of changing the relationships governing the elements engaged in the conflict. It is characterized by a sharp break from the ordinary, shortness of duration, high risk and grave implications for the stability of the system.

A crisis can be due to:

- *Natural causes*: such as earthquakes, volcanoes, floods, tornadoes, natural fires, lightning strikes.

- *Human causes*: such as administrative conflicts, bad management and an accumulation of unsolved problems.

- *Industrial causes*: usually from equipment, apparatus and other technologically related causes.

Flexibility characterizes the way we deal with problems, but when dealing with a crisis, flexibility may mean indecisiveness. In the case of a crisis we cannot change the past, thus we have to act firmly and fast, before it develops into a disaster.

1.1.3 Disaster

The disaster is the result of a crisis or a series of crises. We cannot bear its consequences for a long time, since the life-threatening implications and cost of loss due to a disaster are very high. Three elements can be identified when discussing a disaster:

- The effect of natural (and other) hazards (disasters) such as earthquakes, hurricanes and floods. These are in fact natural agents that transform the vulnerable systems condition into a disaster. So we can consider them as factors causing the disaster.

- The effect of the event causing the disaster on the system, the people and the environment.

- The human activities that increase the effect of the disaster.

1.2 CONTEMPLATING DISASTER

What every operation and maintenance engineer strives to achieve in their work is to insure the system operates without interruption. To do this, they must:

- Maintain the system properly so that faults become a remote possibility.

- Understand the system very well, so that when a fault happens they are ready and able to solve it before it develops into a disaster.

- Think of the worst that can happen to the system, and try to find ways and means that can help rectify the worst possible condition.

- Train themselves on how to cope with disaster, by simulating unthinkable situations and trying to solve them.

1.2.1 Example

Suppose you are maintaining the power-generating facilities in a hospital, the worst that can happen is a mains power supply breakdown in the operating theater during an operation. To safeguard against this scenario, you install an auxiliary power supply unit that is automatically inserted in the circuit when a power failure occurs.

The worst that can happen, the unthinkable, is that the mains power supply is interrupted, and the auxiliary power supply does not operate automatically while a surgeon is operating in the operating theater. This is the unthinkable situation that every O&M engineer should think of when dealing with an anticipated disaster, however remote it might be, because in this case the life of a human being will be at stake.

1.2.2 Why and how

If we go back to the previous example, we find that anticipating how a disaster can happen will help in counteracting its effects or at least in reducing its disastrous effects. This can be done as follows:

- Drawing the suggested connection diagram for the mains power supply and the auxiliary one.

- Performing the 'what if' analysis; this begins by asking ourselves about the worst- case event that can happen, which is the interruption of power in the operating theater.

- Analyzing the reasons that permit this worst case to occur.

- Discussing the reasons, to come up with the critical components that should be protected so that this event can never occur.

This process, called failure mode and effect analysis (FMEA), is discussed fully in Chapter 5.

1.3 ASSUME NOTHING, EXPECT EVERYTHING

In disaster management and faultfinding one has to assume nothing but expect everything. Suppose that you were relaying an outside radio broadcast from a remote stadium and the output audio amplifier suddenly lost power. You immediately switched to the auxiliary amplifier and began to look into the cause of this problem (before the auxiliary broke down and a disaster occurred). Let us suppose that a critical transistor circuit was found to be faulty and you were asked to investigate the fault. When measuring the bias voltage on the emitter, you found it was 0.2 volt, almost okay, so you concluded that the bias was okay and decided to move one step further to check the drive.

If in fact the bias circuit was composed of a resistor and capacitor in parallel, and you were measuring with a low input impedance meter, there is a high probability that the measuring instrument's impedance shunted the capacitor to indicate the voltage drop you measured, while in reality the resistor in the bias circuit was open. If you miss things like this, you can go round in circles for some time.

1.3.1 Fault analysis

Usually the fault you face is a simple one, but most of the time we all expect a complicated fault and act according to our faulty assumption. Thus we end up by looking for a complicated explanation of the fault we 'invented', losing time and money. Equipment faults can usually be divided broadly into:

- Faults that happen to equipment during its installation, testing or acceptance, i.e. on equipment that has not actually worked for some time. This can suggest that the fault may be due to design, manufacturing, faulty components or even faulty wiring.

- Faults that occur after the equipment has worked properly for a long time. In this case the fault may be due to a component failure, software problems or other environmentally related aspects (heat, humidity, vibration, shock, dust, radiation, etc.).

- Equipment failure that happens after the equipment has been serviced or due to mishandling.

1.3.2 Routine procedures

It might be strange to deal with the same fault twice. Faults usually do not repeat themselves; each one is different. Thus, we have to use a general rule to locate certain types of fault.

For recently manufactured equipment the diagnostic software is a blessing; but for most equipment in use, the maintenance engineer has to use the old method of measuring voltages and currents, and analyzing waveform shapes.

The best way to cope with these faults is to be prepared to handle them well before they occur. Here is a simple procedure to follow when you receive a new piece of equipment:

1. Attend, or better participate, during the initial inspection and acceptance testing period.

2. If you cannot attend, study a copy of the inspection and test results.

3. Measure and record the voltages, currents and waveform shapes found at the important test and inspection points suggested by the manufacturer. It is even better to identify some other test points that you deem to be critical in the faultfinding process, then record their readings for later use. Compare them with the acceptance test results and investigate any differences.

4. Later on, if and when the equipment gets faulty, you already have your measurements and wave shapes to help you speed your faultfinding activities, and hopefully prevent a disaster.

Checking the obvious

Always start by checking the obvious things. If you are checking the microwave receiver unit in a repeater station, check if the pilot lamps are still alight, check if the tuning meter is still working, even check if the power supply is connected and the fuses are okay. When you start with the obvious and are sure you have checked it well, you are in a position to decide that more detailed investigation is required. You can then begin to reduce the area of search by performing a few simple tests, which should tell you which section of the receiver is faulty. The receiver block diagram may come in handy, as well as the table of voltages and currents you prepared when the unit was working properly. You will now see if there are any changes which will directly reveal the fault.

Working backwards

The most complicated faults often turn out to be a simple mistake or component failure. The idea of working backwards is to check the output and examine the reasons that prevent the output from developing as it should be. Then we go one step backward to check the stage identified as the probable source for the interruption, and perform the same analysis until we discover the components responsible for the interruption. This process is simplified very much if earlier on we have performed the FMEA procedure for this piece of equipment.

Innovations

Telecommunications equipment is becoming more and more sophisticated at a very rapid pace, new equipment is introduced in the market every day. It is thus very hard for the operation and maintenance personnel to cope with the complexity and diversity of such equipment. Equipment manufacturers have identified such a problem and are now including self-diagnosis facilities.

Such innovative ideas will help a lot in the quick identification and clearing of the fault. It is in the best interests of the service engineer to select equipment that helps in quick fault diagnosis, perhaps test points or a diagnostic computer program.

1.3.3 Types of fault

Faults that occur in ordinary electronic equipment can be broadly divided into six categories:

- *Time-dependent faults*: the symptoms here are simple, the equipment works okay for some time, hours or maybe days, but the fault happens at the end. If the equipment is switched off for a period of time and then switched on again, you can operate the equipment okay, but the failure happens again.

- *Temperature-dependent faults*: the equipment fails or changes its characteristics (such as higher bit error rate) or value (resistors and capacitors) with change in temperature. Yet the temperature rise is within the specified operating limits of the equipment.

- *Intermittent faults*: they are the nightmare of the O&M engineer, since the equipment operates okay for some time, for a period that changes all the time, and then fails. The equipment later operates normally for no apparent reason, then goes faulty again, and so on.

- *Software faults*: these are due to a problem in the software; this does not mean there is a virus, but it is usually due to the operator trying a novel procedure or making a wrong move that was not included in the original software program.

- *Dormant faults*: these do not cause immediate system failure and thus cause no further harm. The problem begins when another fault or combination of stress conditions reveals the fault. A dormant fault may accelerate a system failure if it is not detected and treated before it develops (e.g. a security circuit fault).

- *Other faults*: there are many other faults, depending on the types of equipment; they are dealt with later in the book.

1.4 SHARED RESPONSIBILITY

You, as the O&M engineer, or even as the manager, are responsible for disaster management, but you are not alone.

1.4.1 Role of the disaster management team

1. The prime mover in the team will begin by choosing the team members who will select a team leader/coordinator.

2. The team will begin by preparing the preliminary analysis that will be presented to the management to get their approval and support.

3. After securing the management support, the team will develop an action plan detailing the tasks to be performed and assigning responsibilities to team members.

4. Then the team will develop a budget for the risk analysis process.

5. Finally the team will establish the administrative procedures that will be followed during this assignment.

1.4.2 Role of the company management

Management has a very important role in disaster management; it is a key element in the success of the disaster prevention/recovery plan. Yet management should have an incentive to participate actively in such an endeavor; this incentive can stem from the last audit report, a new law that was issued a short time before, or a recent disaster that cost the company a lot.

To begin to talk to management, a preliminary analysis sheet should be prepared. It should answer the frequently asked questions (FAQs) about the project, such as:

- Why should the company go on with the project?

- What are the foreseen dangers of not doing so?

- How much would it cost the company to initiate the project and how much would it cost it to ignore the consequences?

- What is the time frame that is needed?

- What are the main elements of the disaster prevention/recovery project?

1.4.3   Role of the government

The bureaucratic approach of the government is universal; the *Concise Oxford Dictionary* defines bureaucracy as 'government by central administration'. In a bureaucratic government the term 'red tape', which the bureaucrats use to justify doing nothing about something, is very popular. Bureaucracy and red tape are facts of life that we cannot avoid but must learn how to live with.

The impact of a government regulation on a company can be very hard if the company had no prior information about this regulation and couldn't do anything to try to cope with it. So the key factor is to be well informed about the government's attitudes and trends, study its proposed new regulations and try to put forward some amendments or even devise proper rational arguments that suggest the government's proposals are not in fact needed.

1.5   THE COST OF A DISASTER

The cost of a disaster can be calculated according to the type of disaster and those involved in it. It can be broadly divided into four categories:

- Cost of interruption and/or cessation of service
- Damage and destruction of material goods
- Injury and/or fatalities to personnel
- Effects on the natural and social environment

If we take fire as the cause of the disaster, we can divide the costs as follows.

1.5.1   Direct cost

The direct cost includes the cost of human injuries and fatalities, the cost of equipment damage and the cost of lost production. The cost of fire is a direct cost.

Fire can endanger life and limb, since it can cause burns, which in turn causes people to panic and risk their lives, perhaps by jumping from windows. Fire statistics always show the number of fatalities but they do not show the number of injuries or the cost involved due to injuries that give rise to permanent or temporary incapacities, and the cost borne by the companies due to these injuries.

Fatalities due to fire are very hard to evaluate in monetary terms, which makes it very hard to do the loss calculations. Direct costs amount to about 30% of the total cost of fire.

1.5.2   Consequential cost

Indirect costs

Indirect costs are costs not directly related to fire, such as the cost of protecting the equipment and building from the effects of fire, the cost of installing a firefighting system. These indirect costs amount to about 5% of the total cost of fire.

Cost of insurance

Insurance premiums are a very good indicator of disaster costs. Insurance cost is a major item in the budget for any fire. The cost of insurance amounts to about 15% of the total cost of fire.

Cost of emergency measures

When fire breaks down and the installed system does not cope properly, some other measures should be taken into consideration, perhaps the fire brigade needs to be called. Other measures will be taken to insure the continuity of service, perhaps a backup unit will be put in service to carry the traffic during a fire in the exchange room. The cost of emergency measures amounts to about 15% of the total cost of fire.

Cost of prevention

Measures to prevent the occurrence of fire, or to limit its effects if it happens, are very important in saving the company in case of fire. Putting the telephone exchange backup computer in another room indicates that if fire broke out in the main room, the backup computer would not be affected. Yet we have used a spare room at a certain cost; the cost of prevention amounts to 5% of the cost of fire.

Cost of research

Research and regulations are two forms of passive fire prevention. They help in developing a better understanding of the causes and effects of fire and how to fight it more effectively. The cost of research amounts to about 3% of the cost of fire.

Cost of information

The cost of acquiring the proper information about fire prevention and handling fire situations amounts to about 2% of the cost of fire.

1.5.3 Another way to cost a fire

We can look at the cost of fire in another way, by calculating the losses and expenses involved:

Losses due to fire	
Direct losses	30%
Indirect losses	5%
Human losses	5%
Expenses due to fire	
Fire prevention	30%
Insurance expenses	15%
Emergency services expenses	15%

1.5.4 Some cost calculations

Fire in an exchange

Suppose a fire erupted in the exchange room, and we want to calculate the anticipated loss due to this fire to determine the investment we are going to spend in purchasing a fixed firefighting installation. Table 1.1 can be used for this calculation. It assumes that the cost of fire should not exceed 3% of the cost of the installation when a proper firefighting installation is in place. The cost of repair is detailed. In case of fire, use this table to calculate the actual loss and compare it with the anticipated loss, then evaluate the firefighting system accordingly.

Table 1.1 Anticipated cost of loss due to fire

Item	Initial cost ($1000)	Repair cost ($1000)	Total cost ($1000)
Central office	1 000 000	30 000 (3%)	
Cost of fire (not exceeding)	30 000		
Cost of lost service		5 000	
Employees cost (salaries)		5 000	
Cost of new system			
exchange repair		10 000	
wiring		2 000	
UPS units		2 000	
installation		1 000	
firefighting		1 000	
room construction		2 000	
penalties		2 000	
Total cost of loss			30 000

Table 1.2 Loss per hour due to disaster

Number	Facility or activity	Loss per hour (US$)
1	Brokerage operation	6.5 million
2	Credit card authorization system	2.6 million
3	Automatic teller machine	14 000

Hourly cost of interruption

Table 1.2 illustrates the approximate loss per hour for different activities and facilities. This gives an idea about the loss involved and can be compared with the cost of a disaster management and recovery plan.

1.5.5   Insurance

Some companies fall under the illusion that by insuring their operations they will not have to pay large disaster costs, because the insurance will cover the cost of the disaster. There are several reasons why this is not true:

- The insurance company will investigate the company's disaster avoid-ance policy before they decide on the insurance policy premium.

- When a disaster eventually happens, the insurance will cover the recovery costs but will increase the premium of the new insurance policy. This is the same procedure that happens when you renew your car insurance policy every year. If you have not committed any offences or caused any accidents then your new insurance premium will be reduced by a factor of 10–15%. On the other hand, if you have caused many accidents that the company had to pay for, your new premium will definitely be higher than your old one.

- There are several types of coverage, so you have to examine the policy very carefully to know exactly what your company is covered against.

SUMMARY

A problem that goes on without a solution may develop into a crisis, and crises are usually the cause of the disaster. Disasters can be due to:

- *Natural causes*: such as earthquakes, volcanoes, floods, tornadoes, natural fires, lightning strikes.

- *Human causes*: such as administrative conflicts, bad management and an accumulation of unsolved problems.

- *Industrial causes*: usually from equipment, apparatus and other tech-nologically related causes.

What every O&M engineer strives to achieve in their work is to insure the system operates without interruption. To do this, they must:

- Maintain the system properly so that faults become a remote possibility.

- Understand the system very well, so that when a fault happens they are ready and able to solve it before it develops into a disaster.

- Think of the worst that can happen to the system, and try to find ways and means that can help rectify the worst possible condition.

- Train themselves on coping with disaster, by simulating unthinkable situations and trying to solve them.

Disaster management is a shared responsibility, You as the O&M engineer, or even as the manager, are responsible for disaster management, but you are not alone. There is a role for the disaster management team, another role for the company management, and another role for the government.

The cost of a disaster can be calculated according to the type of disaster and those involved in it. It can be broadly divided into these categories:

- Cost of interruption and/or cessation of service

- Damage and destruction of material goods

- Injury and/or fatalities to personnel

- Effects on the natural and social environment

REVIEW QUESTIONS

1. Define the following terms: problem, crisis, disaster. Give some examples from your work experience for a problem, a disaster and a crisis.

2. Explain how you would think methodically about an unthinkable disaster.

3. What are the types of fault analysis used as a disaster prevention measure?

4. What are the roles of routine, preventive and predictive maintenance in disaster prevention?

5. Compare the responsibilities of the disaster management team with the responsibilities of the overall management.

6. How would you estimate the total cost of a disaster, and what are the different subcosts involved in this estimate?

2

Telecommunications Systems: An Overview

OBJECTIVES

- Answer questions at the end of the chapter.
- Discuss the building blocks of telecommunications networks.
- Present examples of network building blocks at your workplace.
- Describe how telecommunications networks are likely to change in the near future.
- Say how these changes might affect disaster management at your workplace.

2.1 FUNDAMENTAL BUILDING BLOCKS

The need for a telecommunications system began in the very early days. Smoke signals, drums and flags are but a few examples. Yet a telecommunications system as we understand it began when the telegraph was invented, followed by the telephone.

A few years later the telecommunications industry took major leaps forward by utilizing the telephone's single copper line pair to carry more than one telephone message. The developments continued until we began to use fiber-optic cables that carry millions of telephone conversations on one cable. This means a huge reduction in the cost of a single telephone call, which is in the best interest of customers.

The problem here is that the cost of failure became huge. If we assume that a single international telephone call costs one dollar and the fiber-optic cable carries one million calls, the cost of failure will be a million dollars per minute. We must also add the cost of repair, perhaps not much in this case.

Now consider a satellite link carrying this telephone traffic; here the cost of repair is huge since you have to send a mission to the satellite to repair the fault or damage. This might cost about 20 million dollars.

This is why disaster management in telecommunications systems is a very important aspect that must be seriously considered by all telecommunications managers, engineers and technicians.

2.1.1   Telephone, telex, fax and data networks

Any telecommunications system is required to transmit the communications information faithfully from its original point to its destination.

This is done using telecommunications networks which developed from the simple telex machine and the plain old telephone system (POTS) to the present-day situation, to cover several more important services to business and residential customers.

A telecommunications network has three main components:

- *Terminal equipment* interfaces between the network and man or machines, e.g. telephone, telex, fax, computer, TV and radio.

- *Switching equipment* is responsible for setting the transmission medium between the two subscribers.

- *Transmission equipment* provides the medium through which the information is exchanged between the subscribers.

Figure 2.1 shows that the telecommunications system is basically a series system; if one link is broken then the system will fail, unless redundant units are available.

The telecommunications system can also be a line system or a radio system. The building blocks of both systems are illustrated in Figure 2.2.

Figure 2.1   The telecommunications system

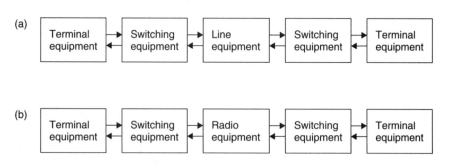

Figure 2.2   Transmission equipment: (a) line, (b) radio

2.1.2   Telecommunications networks

National networks

A telecommunications network is usually composed of the following equipment:

- *Terminal equipment* converts the information into electric signals and exchanges control signals with the network. It can be as simple as the telephone apparatus itself or as complicated as the computer. Terminal equipment interfaces between the user and the machine or network.

- *Switching equipment* is responsible for setting the transmission line between user A (calling subscriber) and user B (called subscriber) so the line can be established. The switching equipment is connected to the transmission line so the link is economical. In the local network, switching equipment can be classified into local exchanges directly connected to the subscriber, and transit exchanges which control traffic between different exchanges.

- *Transmission equipment* is used to connect terminal equipment to exchanges and also to connect exchanges together, quickly and correctly, so that electrical signals are transmitted in the shortest and most economical way between subscribers A and B. Subscriber transmission equipment, which connects terminal equipment to a local exchange, can be metallic cables, optical fibers or radio. Fiber cables are used for the integrated services digital network (ISDN) because it requires a wideband transmission capacity. Transit transmission equipment connects exchanges together in the most economical way; it adopts the optical fiber system, coaxial cable system, microwave system or satellite communication system. Here large amounts of traffic are handled through a single transmission line in the most economical way.

International networks

The international network is the point of interface between different administrations; this is done through the international exchange or the gateway. International networks can be submarine cables, satellite communication links, or sometimes fiber-optic and radio networks.

2.1.3   Network configuration technology

Network configuration technology is used to determine how to form the network by representing the exchanges as points and the transmission lines as lines then calculating the traffic flow in the network. The network can be formed in three basic ways:

- *Mesh*: all exchanges are directly connected to all others.

- *Star*: local exchanges are connected to a transit exchange in a star shape.

- *Composite*: combines the advantages of star and mesh networks.

The network configuration has a very important role to play in the telecommunications system's disaster management process, as we will see later on.

2.1.4 Satellite communications networks

The satellite communications system is composed of three basic units: earth station transmitter, satellite transponder (transmitter/receiver), earth station receiver. Figure 2.3 shows the system configuration. The basic components of a satellite communications earth station are antenna, transmitter (uplink), receiver (downlink), ground communications equipment. Figure 2.4 illustrates the basic components of the uplink and downlink.

Redundancy is used extensively in satellite communications to increase the system's reliability, especially for the satellite system itself, since a major fault in the satellite would require a space mission to the satellite, which would be rather expensive. The reliability of the satellite is around 99.9999%.

2.1.5 Fiber-optic communications networks

A fiber-optic communications system (Figure 2.5) is composed of transmitter, receiver and fiber-optic cable. The transmitter has four subunits:

- The input interface changes analog signals into a digital pulse stream.

- The voltage-to-current convertor interfaces between the input circuit and the light source. The conversion allows the input signal to drive the light source.

- The light source can be either a light-emitting diode (LED) or an injection laser diode (ILD).

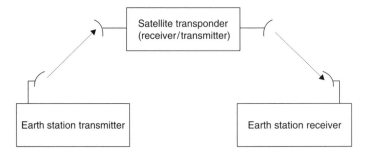

Figure 2.3 The satellite communications system

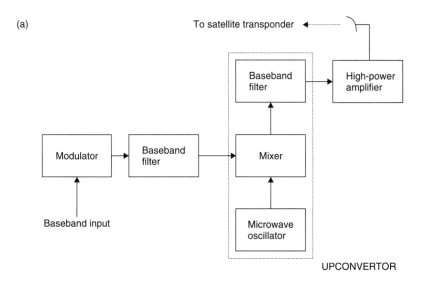

UPCONVERTOR

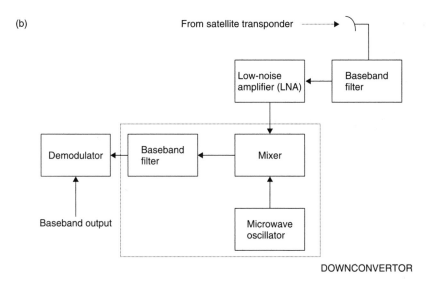

DOWNCONVERTOR

Figure 2.4 Satellite communication: (a) uplink, (b) downlink

- The optical fiber interface couples the light emitted by the source to the fiber-optic cable.

The receiver has three subunits:

- The fiber detector interface couples the fiber to the detector properly; its function is to couple as much light as possible from the fiber-optic cable into the light detector.

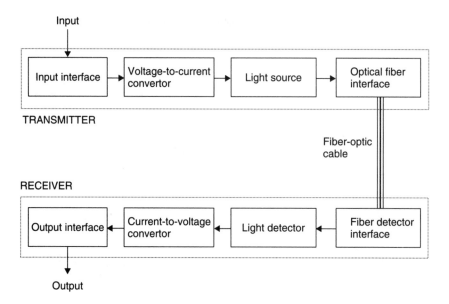

Figure 2.5 The fiber-optic communications system

- The light detector is either a PIN diode or an avalanche photodiode (APD); both convert light energy to current.
- The current-to-voltage convertor produces the output signal.

2.1.6 Computer networks

Local area networks

Local area networks (LANs) are methods of connecting computers, peripherals and communications equipment within a restricted locality such as a building or campus. They operate at very high speed so that data transfer can be effected efficiently.

A local area network consists of the following components:

- *The connection medium* can be a cable such as in the Ethernet developed by Xerox Corporation, or wireless as in the wireless LAN used in mobile communications.
- *The interface* connects the workstation to the cable or wireless equipment.
- *The workstation* can be a computer or other terminal.
- *The server* is a processor that provides shared access to a hard disk (file server) or printer (print server).
- *The gateway* is a separate processor which switches the user into another local or remote network.

The benefits of a LAN network arise out of its distributed processing capabilities. Thus separate workstations on the desk of individual staff members benefit from the combined capabilities of resources stored at each individual workstation.

Wide area networks

A wide area network (WAN) serves the same purpose as a LAN or a metropolitan area network (MAN). The difference between the three networks is in the maximum length of the connecting cable. Thus a LAN has a maximum distance of 10 km, a MAN from 10 to 100 km, and a WAN more than 100 km.

The ISDN network

The switched-circuit technology of the integrated services digital network (ISDN) has taken over the primary means of access to carry speech in a digital rather than analog form, and of course to carry data of any kind. It offers universal connectivity at low cost (due to the huge demand on the service) and with high data rates.

The introduction of broadband ISDN (B-ISDN) has overcome the limitations of narrowband ISDN, allowing the transmission of higher speeds and the introduction of fast packet switching and asynchronous transfer mode (ATM). This has led to high-quality (HQ) broadband video telephony, HQ videoconferencing and high-definition television.

Needless to say, increasing the transmission capabilities means increasing potential losses when a fault occurs.

2.1.7 TV and radio broadcasting networks

The broadcasting system is divided into two main types, sound broadcasting and TV broadcasting (Figure 2.6).

Transmitting stations

Sound broadcasting and television transmitting stations are divided according to the their power, frequency and antenna system. The problem here is that the interruption of service is intolerable from the viewpoint of the administration. The simple solution is to duplicate the system but this is very costly; instead it requires several disaster management techniques to insure continuity of service.

In some countries the continuity of service is important because the management does not want to lose advertisements; in other countries interruption of service is feared and considered as an act of sabotage. Figure 2.7 shows a simplified block diagram for an amplitude-modulated transmitter.

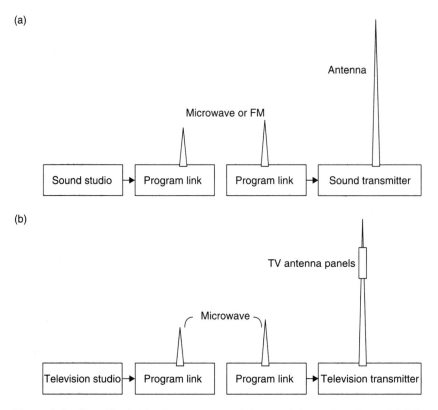

Figure 2.6 Simplified block diagrams: (a) sound broadcasting, (b) television broadcasting

TV and sound broadcasting studios

Sound broadcasting studios are equipped with different audio input equipment, such as microphone, CD player, tape player, disc player and one or two auxiliary inputs. These are connected to a mixing and driver amplifier that feeds the program link which connects the studio to the sound broadcasting transmitter. To insure continued operation there must be redundant equipment in the studio to help in case of technical disaster. Figure 2.8 shows a simplified block diagram for a sound broadcasting studio.

TV studios also have a video section: cameras and their control units, videotape units, synchronization unit, etc. TV studios also feed the program link which connects the TV program to the TV transmitter.

Program links

Digital radio relay links are now used extensively as program links in TV and broadcasting work, as well as in other telecommunications fields. Here are some of the advantages of digital over analog:

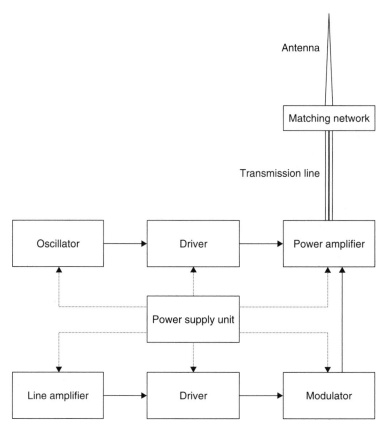

Figure 2.7   AM transmitter: simplified block diagram

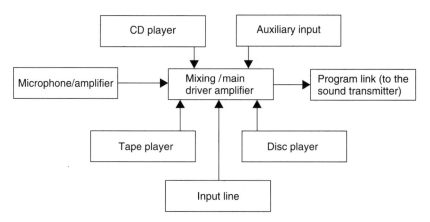

Figure 2.8   Sound broadcasting studio: simplified block diagram

- Easier to interface to digital switching systems, now used extensively in telephone networks.

- Adding multiplexors and demultiplexors does not cause appreciable degradation in the received signal.

- Use of very large scale integration (VLSI) technology increases the reliability, reducing the occurrence probability of an interruption.

- The digital signal is regenerated at each repeater, thus the transmission performance can be assumed independent of the path length, since noise does not accumulate along the transmission path (as in analog transmission).

The digital radio relay link is composed of two terminal stations and a repeater station. The terminal stations interface with the telecommunications network. They also provide signal processing to generate, receive and transmit the required IF signal. The repeater station receives the signal from the microwave receiver, converts it to an IF signal and regenerates it for transmission to the next station. This process eliminates noise and distortion from the signal and produces an almost perfect signal. Figure 2.9 shows a block diagram of a digital radio link.

The effect of fading (degradation of signal strength) can be rectified by frequency diversity and space diversity:

- *Frequency diversity*: two different RF carrier frequencies are modulated with the same IF intelligence, then both RF signals are transmitted to the destination, where both carriers are demodulated and the one that has best performance and yields the best IF signal is selected. Figure 2.10 shows a microwave system using frequency diversity.

- *Space diversity*: the transmitter output is fed to two antennas that are physically separated by an appropriate number of wavelengths. Similarly, at the receiver there are two or more antennas that provide the input signal to the receiver. Figure 2.11 shows the block diagram of a microwave system using space diversity.

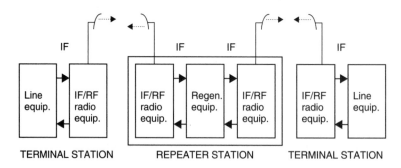

Figure 2.9 Radio relay link

(a)

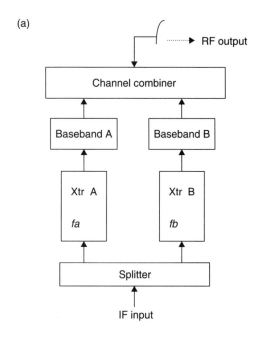

(b)

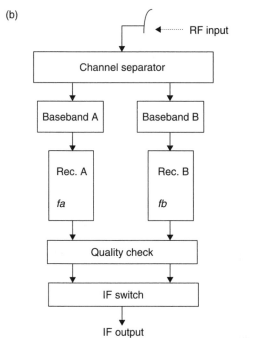

Figure 2.10 Microwave system with frequency diversity: (a) transmitter, (b) receiver

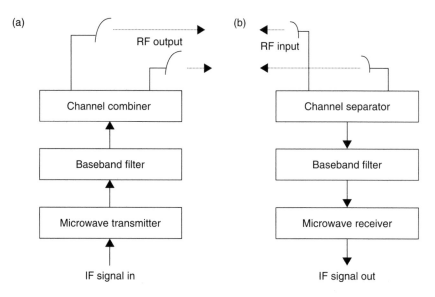

Figure 2.11 Microwave system with space diversity: (a) transmitter, (b) receiver

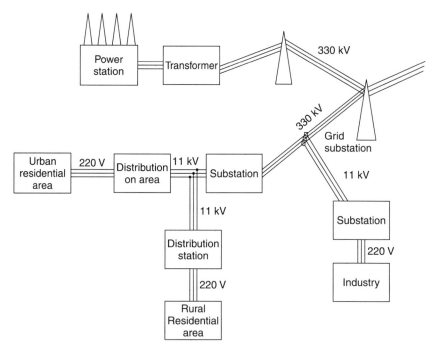

Figure 2.12 AC power generation and distribution

2.1.8   Power distribution networks

Electrical power is generated at far-off power stations, transmitted over long distances on high-voltage transmission lines and supplied at low voltages of 110 V, 220 V or 400 V to loads in urban and rural areas through distribution transformers. Figure 2.12 illustrates the general layout of the AC power supply generation and distribution scheme.

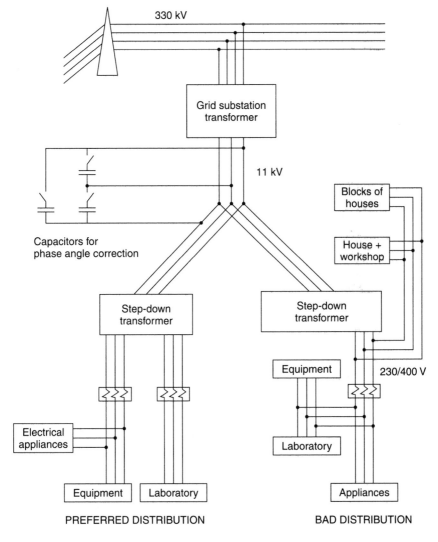

Figure 2.13   Power distribution scheme

High-voltage networks

The high-voltage network carries the electrical power over long distances; that is why the transmission voltage is made high (220 kV) to reduce losses. Figure 2.13 illustrates the preferred way to connect a low-voltage distribution system, compared with another common (but not preferred) method.

Low-voltage networks

Figure 2.14 shows a typical low-voltage distribution system to a company or organization.

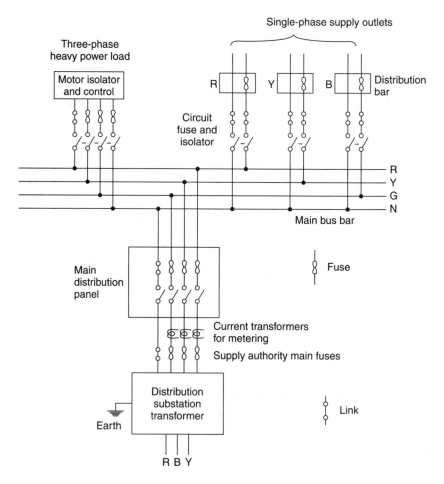

Figure 2.14 Main supply distribution scheme

2.2   HOW A TELECOMMUNICATIONS SYSTEM FAILS

A telecommunications system fails when service is interrupted for a period of time. Consider the cost of interruption to a communications satellite. If we assume this satellite is equipped to handle 250 000 telephone channels for international service and the cost is US$ 1.0 per channel per minute, it means the cost of interruption is US$ 250 000 per minute. If we add to that the cost of loss due to data and TV channel interruption, the figure will become far bigger.

To minimize this loss, the satellite companies insure that a spare satellite is available to carry the service, in case of interruption. This is termed a hot standby, which is only justified when the loss due to service interruptions is enormous.

Alternatively a telecommunications system may fail due to financial problems. I was once contacted by the International Telecommunications Union to visit a developing country and find out why its telecommunications department was incurring such enormous losses.

After thorough investigation I discovered that the majority of the telephone subscribers in this country, who were provided with international dialing, were not paying their international telephone bills, since these telephones were installed in government offices. Moreover, they did not follow any accepted accounting procedures to issue and collect them.

2.2.1   Why some companies fail

Telecommunications companies do not fail instantly, there are warning signs which suggest a disaster may be highly probable. If these early warning signs are ignored, the eventual disaster may be bigger and more dangerous. Thus we have to monitor very carefully these early warning signs and interpret their meaning properly so that quick action can be taken to confront the potential causes.

Bad management

Companies generally fail due to bad management. Bad management can take many forms:

- Autocracy, resistance to change, or lack of control. This situation can be rectified by setting clear and specific goals for the company. Without clear goals, people may procrastinate.

- Inappropriate handling of early warning signs (Figure 2.15):
  — Not detecting early warning signs
  — Early warning signs detected but ignored
  — Early warning signs detected and hidden

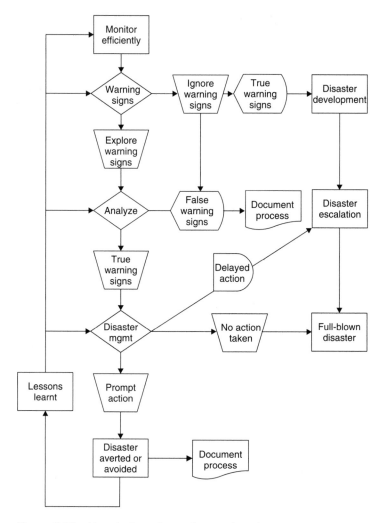

Figure 2.15 How to handle early warning signs

- Decisive action rarely taken when the company is at risk. The managers do not want to take a decision because they:
  — Fear it is not the right one
  — Hope somebody else will take the decision
  — Hope the problem will solve itself in time
  — Procrastinate or act ineffectually

- Routine checkup procedures not implemented. To manage disaster properly, we must first make sure the occurrence probability of a disaster is minimized by performing routine checkups. If they are not performed routinely and the resulting data analyzed properly, then we have a higher probability of a disaster.

Good management

Here are the basic concepts of good management:

- Have a well-defined strategy.
- Implement strict control measures, especially cost control.
- Do not manipulate accounts.
- The board of directors actively participate in the company's operations.
- There is no one-person rule in good management.
- Good management knows of and responds to change.
- Good management recognizes, understands and accepts that the customer is king.
- Good managements do business straightforwardly and honestly.
- The organizational structure of good management meets the people's needs and the customers' needs.

2.2.2 The profile of failure

It is true that we learn our lessons by looking at failures, either in our own organization or in other organizations. The six major factors that characterize the failure profile are:

- Top management that lack good leadership and have no sense of urgency.
- Underestimating accounting information, where managers do not appreciate the importance of the mandatory financial statements, and have poor budgetary controls and costing.
- Management do not respond to change, they continue to produce or offer an obsolete product or service.
- Manipulation of accounts (creative accounting), where management knowingly change the way accounts are presented, to conceal important indicators or warning signs.
- Rapid expansion of the company activities to gain a higher share of the market, but without insuring the appropriate funds.
- Ignoring the economic cycle, with big projects going wrong because of the global economic cycle.

2.2.3 Management's role in preventing failure

Many executives have difficulty making decisions, but the biggest enemy of the manager is to take no decision.

Making decisions

The proper business decision-making process contains the following steps:

1. Make decisions by yourself, because invariably they will be true.

2. Define your objectives as clearly as you can.

3. Search for all the information you require, but ask the proper people.

4. List all the available options; list all the ideas for the project and all those against it.

5. Try to be logical and use common sense. Remove any emotions from your decision-making process.

6. The simpler your method for taking the decision, the better you'll be able to implement it.

7. If you make a wrong decision, admit it, correct it and don't do it again.

Setting goals

According to the saying, **A person going nowhere, normally gets there!** So you can decide properly what to do, and be successful in doing it, you have to go through the following steps:

1. Make a list of all your objectives and the goals you *really* want to achieve; list them as:
   - Long-term goals and short-term goals
   - Tangible goals and intangible goals
   - Business goals and private goals

2. Select the *primary goal* from the list, taking into account that:
   - It must be high enough to be worth the extra effort
   - It must be realistic
   - The maximum time needed to achieve this primary goal must not exceed three months

3. Define the goal in complete detail; the more you understand the goal, the more you are able to achieve it.

4. Gather together the correct information; look for the facts but make sure they are the right facts. Never accept what seems to be a fact before checking its validity.

5. Set a deadline for yourself. Decide exactly when you want this goal to be achieved, but be realistic about it.

6. Carry those goals within you and live with them.

7. Imagine what will happen when you have achieved this goal, to be able to face and confront obstacles during the process.

Planning and implementation

Write your selected goal on a sheet of paper and date it. Then set the cutoff date for the completion of the goal. Decide on what you need to achieve this goal, and put the activities involved in sequence. Set a date for each stage to be completed in sequence:

1. Set the objectives accordingly.

2. Appoint key people that are going to be involved in the project.

3. Allocate responsibilities to each person. These responsibilities should be simple and clear, to insure no misunderstandings.

4. Decide the completion date for each objective then work backwards to obtain the deadline date for each component task.

5. Give each person a copy of the plan.

6. Follow the progress ruthlessly and keep everybody informed.

7. Check the budget frequently.

Delegating authority

Delegation is not just for dividing up work; there are several reasons to delegate:

- Delegation creates a second staff line, so the delegator is no longer indispensable and is free to be promoted when the time comes.

- Lack of delegation may leave personnel with insufficient work, hence they are underutilized and overpaid.

- Delegation challenges people, especially the high performers; without these challenges, talented people may leave.

- Delegation helps people to overcome their physical limitations; they work smarter, not harder.

Why people do not delegate

Here are the acknowledged reasons:

- They think their work is top secret and that nobody else can be responsible for it.

- They think their work is of a very high standard that nobody else can attain.
- They are short of staff to perform the work.
- They have no time to train anybody to do the job.

Here are the unacknowledged reasons:

- They are afraid that delegation will rob them of anything to do.
- They are afraid the delegated person will take away their job.
- They think they can do the job better and faster than anybody else.
- They do not want to lose time in training new company personnel to do the job.

Impact of not delegating

- Locking in our low performers, whom nobody expects anything from.
- Forcing out the high performers.
- We start attracting low performers.

How much we should delegate

- We must delegate some work and this must be meaningful work.
- We should delegate until people start to complain, then we should increase delegation by up to 20% more.

2.2.4 The road to recovery

Managing the road to recovery can be achieved in several steps:

1. Identify causes and properly interpret the early symptoms of failure. To prepare a strategy for recovery, we have to perform a diagnostic to search for the real reason behind what has gone wrong. We have to check and double-check the early warning signs that were overlooked.

2. Adopt a better approach to problem understanding, formulation and solution. Concentrate on the real *cause* not the symptom.

3. Set a strategy for recovery. The strategy for recovery must follow certain rules and procedures:
   (a) Define a strategy for the organization.
   (b) Implement strict financial and cost controls.
   (c) The board of directors must participate actively and get fully involved in the operation of the organization.

(d) Teamwork must be the motto of the management.

(e) The board of directors must be broadminded, and they have to respond quickly to any change in the market.

(f) Accounts should be performed properly; they should reflect the real situation of the organization and they mustn't be manipulated.

4. Prepare an action plan to implement the adopted recovery strategy.

5. Involve all inside and outside the company. Involving the employees in the plan is understandable, but what about the outsiders? Well, let us consider a bank manager who has lent the company funds. They must have a say in the plan because they are taking a risk by lending the company money.

6. Establish an efficient process for data collection and analysis. Data can be internal, within the organization, or it can be external. Many items of data come from formal sources such as company transactions, reports or fact sheets. Other items come from informal sources such as word of mouth, gossip, opinions and personal observations.

Learn to recognize what is a fact. Facts can be objectively verified and proven to be true by several means. But distinguish between a fact and what we perceive as reality. Advertisements may contain all kinds of fiction but they are often presented as though they are facts. Try to be honest about your feelings and aware of how they influence your decisions. Many statements we take as facts are actually questions of probability. Do not deny evidence because it may hurt someone's feelings or it flies in the face of a widely held opinion. Do not rely on your senses when they are inadequate for the task; use the appropriate test equipment, e.g. an X-ray machine.

Identifying facts

Inferences

To 'infer' means to bring in explanations (imagined or reasoned), then try to link known facts with missing ones, or to carry us to more facts and conclusions. To 'infer' means to conclude, guess or speculate; it does not mean to imply, hint or suggest. When we infer, we make a guess in order to find explanations that can put the known and the unknown together. When we make inferences, we draw a conclusion by reasoning from evidence. It is important to distinguish inferences from facts, and to draw inferences from careful observations.

Assumptions

To 'assume' is to take for granted. When we make an assumption, we believe in something — we buy it. There are four types of assumption:

- *Unconscious assumptions* are beliefs, values or ideas that are not consciously recognized or expressed.

- *Conscious assumptions* are used as a kind of creative strategy; we make a conscious assumption and use it to guide us to new information.

- *Warranted assumptions* are reasonable assumptions. If you buy a carton of milk dated for use within the week, you can make a warranted assumption it will not be sour when you open it.

- *Unwarranted assumptions* are unreasonable assumptions. They are based on ignorance or lack of awareness.

Opinions

Opinions can be divided into four categories:

- *Judgment*: based on personal or collective codes of value.

- *Advice*: you should do this and not that.

- *Generalization*: all the equipment is unreliable, for example.

- *Personal taste or sentiment*: I like this but I don't like that.

Evaluations

To 'evaluate' is to determine or fix the value or worth of something; it is also to examine or judge, appraise or estimate. This brings up the problem of premature evaluations, since we may deceive ourselves into thinking our evaluations are based on our accumulated wisdom and experience.

Evaluations are not facts, and they are influenced by expectations. Thus, skilled use of evaluations is essential for identifying the real facts, as much as discovering evaluations covered by connotative words. (I do not like the behavior of this machine, meaning it is going to fail.)

Viewpoints

There are conscious and unconscious viewpoints.

Dealing with data

After we have collected the data, we need to use it efficiently; consider the following example. Suppose you monitored the maintenance period in minutes and recorded the following 50 results.

40	58	43	45	63	83	75	66	93	92	71
52	55	64	37	62	72	97	76	75	75	64
48	39	69	71	46	59	68	64	67	41	54
30	53	48	83	33	50	63	86	74	51	72
87	37	57	59	65	63					

From such raw data we can deduce almost nothing to help us reveal hidden patterns. One method is to calculate the arithmetic mean. In this case it is the sum of all the measurements divided by 50: 3095/50 = 61.9 =~ 62 minutes. Is this value acceptable? We check the manufacturer's specifications then we do a standard deviation calculation.

The standard deviation is a relation between the average time (mean time) and the actual maintenance time:

$$\sigma = \left(\frac{1}{n-1} \sum_i (\text{time } i - \text{mean time})^2 \right)^{1/2}$$

$$= (13\,013/49)^{1/2} = \sim16 \text{ minutes}$$

Thus, according to the well-known 3σ rule, we have this new information:

- 68% of the measurements fall within the interval mean $\pm\sigma = (62 - 16)$ to $(62 + 16) = 46$ to 78.

- 95% of the measurements fall within the interval mean $\pm2\sigma = 30$ to 94.

- 99% of the measurements fall within the interval mean $\pm3\sigma = 14$ to 110.

All this can be performed in seconds with an ordinary hand-held scientific calculator to give the O&M engineer an insight into why the maintenance period varies on this piece of equipment. This may help in preventing a hidden disaster.

Using PERT

The project evaluation and review technique (PERT) was developed in the United States during 1959 to plan and control development progress on the Polaris Fleet Ballistic Missile Program. Yet it is now widely used as a powerful tool in optimizing project implementation and control. It is very useful in disaster management since it identifies all the critical activities within a project that can cause delay in its implementation or increase costs if a delay occurs. It is a very useful tool for the disaster manager.

All project activities are sorted and the most probable time to implement each of them is identified. The cost of performing each activity is calculated on the assumption that the activity will not be retarded or advanced. The extra cost needed to complete each activity one or more days earlier is also calculated. A network is drawn to simulate the project activities and the optimum time for project completion is calculated.

Installing a microwave station

Referring to Figure 2.16, we will assume that the needed activities are installing the reinforced concrete column at site B, installing the towers

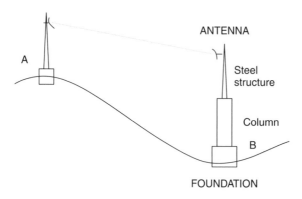

ANTENNA

Steel
structure

Column

B

FOUNDATION

Figure 2.16 Site layout

and antennas at sites A and B, then leveling the two dishes together. The first step is to estimate the most likely time for each activity to be completed, and to determine the activities that must be performed first and those that must follow (Table 2.1). From Table 2.1 we know that the total cost of the project is US$ 1430, but we do not know when the project will be completed.

The PERT network

The PERT network is drawn to reflect the logical performance of the activities. Decide the sequence of activities; we have to survey the site before digging the foundation, and we have to erect the tower before we

Table 2.1 PERT activities

Event	Cost (US$)	Duration (days)	Activity	Activity description
1	50	1	0–1	Surveying the site
2	100	4	1–2	Digging the foundation for column B
3	300	3	2–3	Laying the foundation for column B
4	500	2	3–4	Installing column B
5	20	0.5	4–7	Level adjustment
6	50	2	1–5	Digging the foundation for column A
7	100	1	5–6	Laying the foundation for column A
8	20	0.5	6–7	Level adjustment
9	100	2	7–8	Erection of the B tower structure
10	50	1	8–9	Erection of the B antenna structure
11	20	0.5	9–12	Level adjustment in site B
12	50	1	7–10	Erection of the A tower structure
13	50	1	10–11	Erection of the A antenna structure
14	20	0.5	11–12	Level adjustment in site A
Total	1430	?		

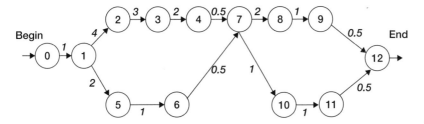

Figure 2.17 PERT network

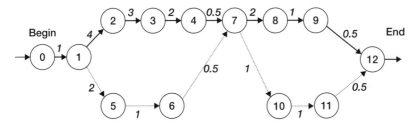

Figure 2.18 The critical path 0-1-2-3-4-7-8-9-12

install the antennas, and so on. We also identify what activities can be performed simultaneously. The resulting network is shown in Figure 2.17.

To calculate the total duration of the project, we have to determine the critical path. The critical path is composed of several activities in series; a delay of one day in any of the activities will delay the whole project by one day. Similarly, an advance of one day in any of the activities will advance the whole project by one day. This is illustrated in Figure 2.18.

The critical path network

The critical path is composed of the following activities, 0–1, 1–2, 2–3, 3–4, 4–7, 7–8, 8–9, 9–12. The reason is that the loop composed of events 1, 5, 6, 7 takes 3.5 days to complete whereas the loop composed of events 1, 2, 3, 4, 7 takes 9.5 days. Now since event 7 cannot be completed unless both loops are completed, the controlling factor will be within the second loop — the longer implementation loop. If the first loop is completed on time in 3.5 days, event 7 will not be completed unless the second loop is completed 6 days later. Thus the project will take 14 days to be completed and will cost US$ 1430.

Optimizing the network

Optimization aims to arrive at the shortest possible completion time and at a reasonable cost. To do this, we have to prepare a table containing the cost of performing each activity and the cost per day for accelerating

Table 2.2 The PERT time/cost analysis

Event	Cost (US$)	Duration (days)	Activity	Accelerated time (days)	Acceleration cost per day (US$)	Added costs due to acceleration for different solutions (US$)			
						1	2	3	4
1	50	1*	0–1	1	—	—	—	—	—
2	100	4	1–2	2	50	100	100	100	100
3	300	3	2–3	2	100	200	200	200	200
4	500	2	3–4	1	350	350	—	—	—
5	20	0.5*	4–7	0.5	—	—	—	—	—
6	50	2	1–5	1	50	50	50	50	—
7	100	1	5–6	0.5	200	100	100	—	—
8	20	0.5*	6–7	0.5	—	—	—	—	—
9	100	2	7–8	1	100	100	100	100	100
10	50	1	8–9	0.5	100	50	50	50	50
11	20	0.5*	9–12	0.5	—	—	—	—	—
12	50	1*	7–10	1	—	—	—	—	—
13	50	1	10–11	0.5	—	—	—	—	—
14	20	0.5*	11–12	0.5	—	—	—	—	—
Total	1430	?							

* Indicates an activity that cannot be accelerated.

each activity (Table 2.2). Some activities cannot be accelerated and they are marked with an asterisk.

Solution 1: Accelerate all activities

- The critical path is 0–1–2–3–4–7–10–11–12 (Figure 2.19)
- The cost is US$ 2380 (66% increase in cost)
- The duration is 9 days

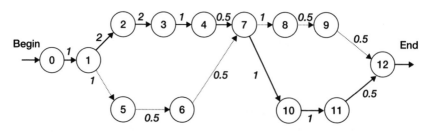

Figure 2.19 Solution 1

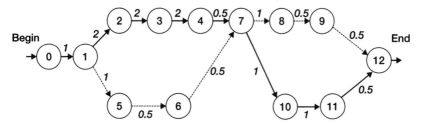

Figure 2.20 Solution 2

Solution 2: Accelerate all except activity 3–4

This was chosen because accelerating activity 3–4 by one day will cost the project US$ 350, which is excessive.

- The critical path is 0–1–2–3–4–7–10–11–12 (Figure 2.20)
- The cost is US$ 2030 (42% increase in cost)
- The duration is 10 days

Solution 3: Accelerate all except activities 3–4 and 5–6

This was chosen since activity 5–6 is not on the critical path.

- The critical path is 0–1–2–3–4–7–10–11–12 (Figure 2.21)
- The cost is US$ 1930 (35% increase in cost)
- The duration is 10 days

Solution 4: Accelerate all except activities 3–4, 5–6 and 1–5

This was chosen since activity 1–5 is not on the critical path.

- The critical path is 0–1–2–3–4–7–10–11–12 (Figure 2.22)
- The cost is US$ 1880
- The duration is 10 days

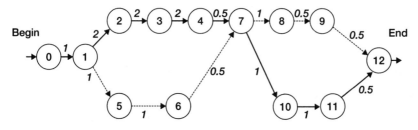

Figure 2.21 Solution 3

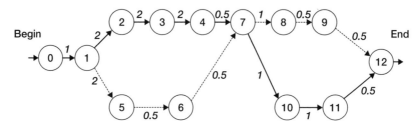

Figure 2.22   Solution 4

We have arrived at the shortest implementation period (10 days) with the lowest cost (US$ 1880), i.e. with only 31% increase over the original budget. If we look at the cost per day in Solution 1 and Solution 4, we obtain the following comparison:

$$\text{Solution 1:} \quad 2380/9 \ = \text{US\$ 264 per day}$$

$$\text{Solution 4:} \quad 1880/10 = \text{US\$ 180 per day}$$

Solution 4 has a lower cost per day. This process can be carried out on a personal computer.

2.2.5   The role of the manager

A company fails when the real profits before tax, measured at constant prices, decline for three or more successive years. The managing director or chief executive officer is usually the one to be blamed.

The common tendency is to blame others, since some managers are good at finding excuses when things go wrong, and none of them is willing to accept that failures come from their bad management. This is clear, since while some companies suffer and others fail, there are yet others who manage to prosper in the same field.

What makes a good manager

- Moving between the past, present and future of the company.
- Demonstrating a clear and convincing route to the company's objectives.
- Setting a good example.
- Being a guide, counselor and mentor to others in the company.
- Putting forward new ideas and obtaining their acceptance.
- Learning from success and failure, their own and other people's.
- Foreseeing technological change and setting strategies accordingly.

- Being a self-achiever, but also achieving results through others.

- Knowing answers to the following questions about the company's future:
  — Where are we now?
  — Why did we get here?
  — Where must we be tomorrow?
  — How do we get there?
  — When should we be there?

2.2.6 Company turnaround

Appendix F deals with the rescue of a major engineering company; it looks at the rescue strategy, the results it achieved and the lessons that were learned.

2.3 THE FUTURE OF TELECOMMUNICATIONS

2.3.1 Introduction

The new telecommunications system of the twenty-first century will no longer concentrate on the provision of standard basic services to end users and businesses, it will depend on how fast they can provide customers with unique packages of applications and network services, to meet creative consumers and businesses. This will depend on the reliability, quality, speed and advanced performance of the telecommunications networks. It will also depend on how creative the telecommunications companies can be in offering their customers more new services.

To meet these demands the companies will certainly use a new generation of fully digital high-speed optical networking in the switching, transmission and outside plant. The optical fiber's wide bandwidth and network flexibility are needed to handle the new network protocol capable of merging each customer's telecommunications requirements into a single item at their fingertips.

2.3.2 The network of the future

Telecommunications advances in the coming decade will be much faster than we have ever anticipated. Several products give very clear signs of this:

- A new class of silicon technology called system-on-a-chip is being developed

- Photonics technology is doubling fiber-optic capacity every year. In 15 years it is anticipated that a single fiber will carry 1000 wavelengths, with a total capacity of a quadrillion bits per second. Dense wavelength division multiplexing (DWDM) will give a limitless increase in transmission capacity.

- Wireless radio transmission is gaining ground very quickly, making it easier to bring high bandwidth to places unreachable by fiber optics.

- Software will enable us to do whatever we want at any time, anywhere, with any device and using any communications medium. It will enable the network operator to tailor services to individual customer needs and preferences.

- Soft switches will provide quick and faultless transfer of traffic between Internet protocol (IP) and circuit networks, creating new communications services.

Here is a general outline of the new network:

- A new network that links the existing networks, protocols and devices together.

- A high-reliability network that guarantees high-quality service.

- A very wide bandwidth network to cater immediately (without any waiting) for all customer requirements.

- Advanced software to link all these services together and to create more advanced services.

- Network flexibility that quickly responds to market changes.

- Reduced costs to attract a prime share of the market.

2.3.3 The IMT 2000 system

The main concern of manufacturers and carriers used to be providing increased bandwidth to cope with the extraordinarily fast spread of the Internet. It is estimated that Internet subscribers will reach one billion in 2003 and will outnumber the fixed telephone service by 2006. Now they have three new concerns:

- To allow customers to obtain all the information they need at any time and anywhere in the world.

- To eliminate any problems due to differences in access or in core network technologies.

- To insure that all e-commerce transactions are secure.

The only way these consumer services can be realized is through wireless communications. The introduction of the International Mobile Telecommunications IMT 2000 system or third-generation wireless telecommunications will overcome some of the major incompatibilities

Table 2.3 Wireless communications systems

Area	Analog system	Digital system
Americas	AMPS, CTI	D-AMPS, CDMA,PACS
Europe and Middle East	NMT, TACS	GSM, DECT
Japan	JTAC	PDC, CDMA
Asia	NMT, TACS, AMPS	GSM, CDMA

that exist with present-day systems. Table 2.3 lists some of the incompatible wireless systems that currently exist.

The third-generation IMT 2000 will have the following characteristics:

- Increased use of digital technology
- A single flexible radio interface for multiple operating environments
- A common global frequency band
- Increased transmission speed (>2 MB/s)
- Improved spectrum utilization and overall costs
- Efficient interworking with earlier systems
- International mobility roaming

It is anticipated that the mobile multimedia communications terminal will be a foldable palmtop with the following facilities:

- Palmtop standard capabilities
- Video camera, video on demand and conferencing
- Multimedia document transfer
- Several multimedia applications
- Handwriting recognition
- WWW, e-mail, e-commerce and fax
- TV, radio and music on demand

The main problem with the implementation of the IMT 2000 system is in finding a vacant global frequency slot in the overcrowded spectrum. This problem was partially solved during the World Administrative Radio Conference (WARC) organized by the ITU. It is hoped that a complete solution will be arrived at during subsequent conferences. These advantages of the IMT 2000 system will no doubt have a positive impact on the importance and criticality of disaster management and recovery planning.

2.4 NEW STRATEGIES FOR DISASTER MANAGEMENT

When we have very wide bandwidth digital networks carrying vast amounts of data, any interruption will mean enormous financial losses. Disaster management will be more important than ever. Proper plans must be drawn up then implemented correctly.

Technicians and engineers cannot perform their duties within the plan unless they are well trained before the disaster erupts. This means that the training component in disaster management will become paramount, because we depend on human beings whose reaction is sometimes unpredictable.

The increasing sophistication of network technology makes it essential to employ highly competent engineers and technicians. The network of the future has no place for the skilled, let alone the semiskilled, but only the highly skilled. Entry requirements will have to be changed to reflect this need and maybe educational syllabuses will have to be altered.

SUMMARY

After a general overview of telecommunications systems, we discussed how a telecommunications system might fail, outlining the failure profile and the road to recovery. We also looked at how managers can influence the success of a whole company.

Looking to the future of telecommunications systems, we gave a short account of IMT 2000 and how increasing sophistication would increase the need for disaster management.

REVIEW QUESTIONS

1. Define the following terms: LAN, WAN, MAN, ISDN.

2. Explain how a telecommunications system is organized.

3. Discuss the block diagram of one telecommunications system, outlining the function of each block.

4. Explain how a telecommunications organization might fail. What are the symptoms of failure that you would look for?

5. What are the dangers of company failures? Give some examples.

6. Explain how you might prepare a PERT program for a maintenance process at your workplace. How might this help your company overall?

3

Some Common Interruptions

OBJECTIVES

- Answer questions at the end of the chapter.

- Discuss interruptions, breakdowns and disasters in your organization.

- Look at the financial impact of events in your workplace.

- Describe some events you might expect at your workplace.

- Say how an event might affect your disaster management strategy.

3.1 BREAKDOWNS AND DISASTERS

The definition of a minor or major disaster can be very tricky; it very much depends on the function of the job and the responsibilities it entails. Yet the definition can also depend on many variables, and the choice of these variables will again depend on the type of work involved.

Thus, we arrive at a common starting ground. Let us take the *time of interruption* as the main parameter we are concerned with. In this case an interruption of a few minutes might be considered a minor interruption causing minor disaster (if any), whereas an interruption of half an hour (or more) can be considered as a major breakdown causing major disaster.

If we apply this definition to the electrical power supply for a residential area, it seems okay since not much is likely to happen if electricity is interrupted for a few minutes. Also, a telecommunications exchange will not suffer any harm since the emergency power supply unit will operate and the consumer will hardly notice a difference in the quality of service offered. Yet if this electrical power interruption were to happen to a high-power broadcasting station, nothing could be done about it because the emergency power generator would not be able to handle the full load.

The matter would be aggravated if there were an important political address or program being transmitted at the moment of interruption. Here an interruption of one minute would be regarded as a major disaster.

If we take the *scope of the interruption* as the parameter, we might assume that an interruption in a part of the system could be considered as a minor breakdown causing minor disaster. This concept can be applied to certain telecommunications systems where there are many alternative routes (parallel systems), but in a series system it fails. Let us take the example of a TV studio where a video camera breaks down during the recording of a show. In this case the director, who usually uses three cameras at the same time, switches to the second camera and the viewer doesn't notice any interruption. But on the TV transmitter side, a series system by definition, if the first RF stage breaks down then the transmission stops.

So it is up to the O&M engineer to define what might be considered a minor interruption and what should be treated as a major disaster.

3.2   THE HEALTH OF AN ORGANIZATION

The role played by interruptions, breakdowns and disasters in the life of a telecommunications organization cannot be overemphasized. The health of the telecommunications organization depends on its ability to supply continuous, high-quality service to its customers. If this is not possible, the customers will look to other telecommunications service providers to satisfy their needs.

Besides losing customers, the telecommunications organization which suffers from breakdowns will be obliged to recruit more technical staff to solve these problems. Those staff must be well trained on the sophisticated telecommunications equipment, and this will be an added expenditure.

The organization will also be obliged to pay compensation to customers whose interests were affected by the service interruption; this is another added expense.

The high cost of spare parts — the traveling wave tube (TWT) at the earth station is a very expensive item — is another factor that adds to the losses of the organization.

The cost of keeping a large inventory of spare parts is another element in the loss chain. In healthy situations there is no need for a large inventory.

3.3   OPERATIONAL AND FINANCIAL IMPACTS

The operational impact of the disaster has been indicated earlier, and will be dealt with fully later on. But to deal with the financial impact of the disaster on the health of the organization, one has to look into accounts books and understand how the accountants prepare their financial statements. This will help to explain how interruptions affect the financial health of the organization.

Here are the main elements to consider when compiling the financial calculations for a telecommunications organization:

- *Capital costs* include the cost of the loop, the buildings, the exchanges, the transmission equipment, the junction routes and the trunk routes.

- *Operating costs* include the maintenance cost and the cost of fault-finding and repair in the loop, the buildings, the exchanges, the transmission equipment, the junction routes and the trunk routes.

- *Depreciation cost* is the amount of funds reserved annually so that at the end of the useful life of the equipment a new item can be purchased with that amount of reserved funds. Usually buildings are depreciated for 50 years, ducts for 40 years, cables for 20 years, exchanges and transmission equipment for 15 years, although they might become obsolete long before that date, due to the pace of technological advancement.

- *Installation costs* are the costs of installing the equipment.

- *Support* is the cost of all other work related to the above activities

- *Headquarters overhead* is the cost of the people who are not directly involved in the above activities; it also includes the cost of the accountants and all support staff.

- *Revenue* includes all the funds the organization gets in return for the service given, such as the local, trunk and long-distance traffic, the connection fee for telephones, and any other revenue gained from selling the different services offered.

Knowing the financial details of the telecommunications organization enables us to prepare the three main financial statements required to insure the organization is operating properly.

Profit and loss statement

The profit and loss statement shows the profit retained by the organization after all expenses have been catered for. It can be explained as follows:

- The total revenue is used, in part, to pay for the depreciation, the overheads (usually taken as 10% of the operating cost), and the operating costs. The remainder is termed the operating profit or earnings before interest and taxes (EBIT).

- The telecommunications organization borrows funds from the bank so it can implement its projects, thus the interest on this loan should be paid from the operating profit, and the remainder is termed the profit before tax.

- The telecommunications organization also has to pay taxes to the government; these taxes are deducted from the profit before tax, and the remainder is termed shareholder profit.

- After distributing the shareholder dividends (profits), the shares of the employees and the board of directors, the remainder is termed the retained profit, which the telecommunications company uses in funding its expansion program (instead of borrowing from the bank).

Balance sheet

The balance sheet is a financial statement showing that the 'source of funds' of the telecommunications organization is equal to its fixed and current assets minus its current liabilities. This can be calculated as follows:

- Fixed assets = capital costs minus depreciation

- Current assets = deposit in the bank, cash at hand, inventory, portfolio investment of short-term nature, plus funds owed by debtors (which preferably should not be more than 20% of the revenue)

- Current liabilities = tax to be paid on profits plus dividends to shareholders

- Net assets = fixed and current assets plus current liabilities

- Net assets are financed by loans from the bank plus shareholder capital plus retained profit

Cash flow

The cash flow is a financial statement that helps insure there will be cash in the bank to pay the telecommunications company's liabilities as and when they are due, so that no extra interest on loans is required. This can be calculated as follows:

- The sources of funds in the bank are the loan taken from the bank plus the shareholder capital put in the bank. They are used in part to pay the necessary costs.

- The first payment will be for the fixed assets, which are composed of the capital costs minus the depreciation.

- The second payment will be for the interest on borrowed funds, and payment of the principal too.

- The third payment is for the operating costs.

- The fourth payment is for the overheads.

- The money in the bank will be supplemented monthly by the cash revenue realized by the telecommunications organization (minus the funds owed by the debtors).

- Thus the cash flow will be healthy if there is enough money in the bank whenever an expense has to be paid.

Suppose a disaster happened, the service was interrupted for a long period, spare parts were needed, revenue was lost, customer claims due to interruption of service were filed against the telecommunications organization, and the organization lost money. All these events and related expenditures will reflect on the cash flow sheet (and the other financial statements) and will result in more funds to be spent than already contracted for, resulting in a gap in the cash flow. This means the organization will have to borrow more money and pay more interest, which will reflect negatively on the profit and loss account of the organization. Chapter 8 deals with this topic in greater detail.

3.4   A FEW SPECIFIC TOPICS

3.4.1   Basics of faultfinding

Circuits and test readings

When a component in a circuit fails, certain failure symptoms prevail. Very often these symptoms are unique to that specific fault; they will change the circuit operation and vary the level of voltage, current or wave shape of the resulting signal. Thus test readings taken when the circuit was functioning properly, come in very handy during a failure.

The block diagram of the faulty equipment is very important in localizing the stage where the fault is expected to be. Sometimes the block diagram is not provided by the manufacturer, so it is the duty of the faultfinding team to construct it from the manufacturer's detailed circuit diagram. A sample block diagram for an RF signal generator is given in Figure 3.1.

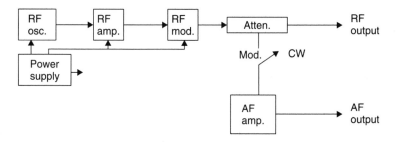

Figure 3.1   RF signal generator

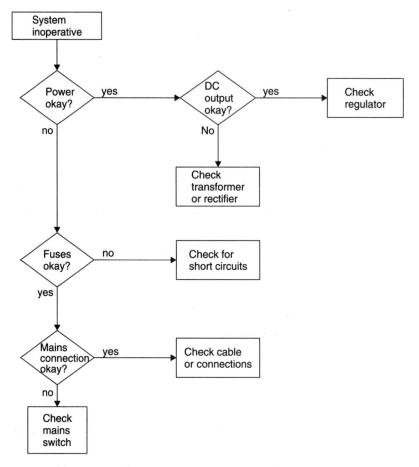

Figure 3.2 Fault detection flowchart

A flowchart for fault diagnosis is another important tool; it can also be prepared by the faultfinding team if it is not already provided by the manufacturer. Figure 3.2 shows a sample flowchart to detect a fault in a regulated power supply unit.

Analog or digital multimeters take voltage, current and resistance measurements, but waveforms and phase shift are taken by a general-purpose oscilloscope.

Basic components, common faults

Resistors

Resistors are devices that provide a specific amount of resistance to the current flow in a circuit. They can be either wirewound or carbon composition. Fixed or variable, the resistance depends on:

- The material used (in composition resistors)
- The length of the wire used (in wirewound resistors)
- The cross-sectional area of the conductor
- Temperature (the higher the temperature, the higher the resistance)

Resistors fail in three ways:

- By becoming open circuit
- By changing their value
- By becoming short circuit

These failures can be detected by measurement using a multimeter set on the resistance measurement scale.

Capacitors

A capacitor is a passive component that stores electrical energy. It is a device that opposes the change in circuit voltage. In its simplest form, it is composed of two plates isolated by a dielectric material. The capacity of a capacitor depends on:

- Area of the plates
- Distance between the plates
- Material used as dielectric

There are different types of capacitors such as air, mica, paper, ceramic, electrolytic, tantalum. Capacitors can fail in three ways.

- Becoming open circuit
- Becoming short circuit
- By changing their value

These failures can be detected by measurement using a multimeter set on the capacitance measurement scale, or by using the LC bridge.

Inductors

Inductance is the ability of a coil to oppose any change in the circuit current. The inductor is composed of a coil (usually copper wire) wound around a tube. The tube can be filled with air or by a core made of a material that has certain permeability to increase the inductance. The inductance of a coil is affected by:

- The number of turns, N ($L \propto N^2$)
- The permeability of the core material

- The cross-sectional area
- The length of the coil

When two coils are brought near each other, mutual inductance develops. Factors affecting the influence of mutual inductance are:

- Physical size of the two coils
- Distance between the two coils
- Permeability of the cores (ferrite has very high permeability)
- Number of turns in each coil
- Distance between the axes of the two coils

Inductors fail in several ways:

- By becoming open circuit
- By becoming short circuit
- By changing their value

These failures can be detected by measurement using a multimeter set on the resistance measurement scale, or by using an *LC* bridge.

Common active components

Semiconductor diodes

Semiconductor diodes are made of germanium or silicon crystals, which in their pure form cannot function as a semiconductor device. Yet with the addition of some impurities, the crystals can be made to conduct a current.

When arsenic is added to germanium it will transform it into n-type germanium, because some free electrons are relatively free in the crystal structure. When indium is added to germanium it will transform it into p-type germanium, because some positive carriers (holes) are developed.

A diode is composed of the two types of crystal bonded together. The diode will be conducting in one direction and nonconducting in the other direction. Diodes are tested using a multimeter and the method of testing depends on their one-way conduction property.

Voltage regulator diodes

When the voltage is reversed on a diode, high current will pass and the diode will be destroyed. This property is used in the reverse-bias diode found in voltage regulators and other applications, such as:

- Avalanche breakdown
- Zener effect

- Voltage regulation
- Switching

Transistors

A transistor is a device that has more impedance in the input than in the output (or the other way around depending on its connection in the circuit). Due to this difference in impedance between input and output, the transistor is able to amplify signals. This characteristic of impedance difference is used in its testing, which can be performed by a transistor tester if the transistor can be checked outside the printed board, or by a multitester on the ohm scale if it is not possible to take the transistor out of the printed board.

Integrated circuits

The integrated circuit is a single chip made of semiconductor material. On this chip are deposited thousands of transistors, diodes, resistors and capacitors. The chip is tested by checking the input/output pulses.

Lasers

A laser is a device that emits microwave energy in a very narrow, highly intense beam of light. It can be focused over long distances. The term 'laser' stands for light amplification by stimulated emission of radiation. Lasers are used extensively in telecommunications applications, especially with the introduction of fiber optics

Measuring instruments and test methods

Meters

When an electronic equipment fails, the repair technician must obtain more information about the symptoms of the failure. Here the multimeter comes in very handy because it can measure voltages, currents, impedance and some other parameters. This information plus some additional data about the circuit performance, like a distorted output or overheating of a component, will be sufficient for the service technician to begin faultfinding work. So a general-purpose multirange meter is essential for faultfinding work.

Oscilloscopes

The cathode-ray oscillosocope (CRO) is one of the most useful faultfinding instruments, since we can use it to measure AC and DC voltages and currents, phase angle and many other parameters. For a CRO to be useful in analog and digital work it must be wideband; but the greater the bandwidth, the higher the cost. However, the vast capabilities of a wideband CRO usually justify the extra expense.

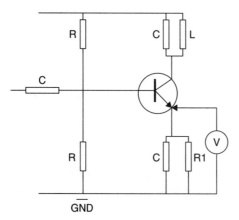

Figure 3.3 One-transistor amplifier stage

Simple testing

When checks indicate a circuit is faulty, then the multimeter comes in handy. Its use goes beyond measuring voltage, current, impedance or other parameters, and extends to analyzing the measurements so the fault can be detected and the real cause identified.

It is important to test the faulty component to verify the type of fault and to collect data on component failure, which will be used later on to upgrade the maintenance plan, in order to increase the reliability of the equipment being maintained.

To illustrate the importance of properly interpreting the measurements, consider the simple one-transistor circuit in Figure 3.3. A voltage drop was detected on the bias resistor R1. One might conclude that the transistor is okay, yet the value of the voltage drop was less than the anticipated figure, which means that the drop was due to the internal resistance of the measuring unit. The transistor is therefore expected to be faulty, which was the case.

Faultfinding electronics

When repairing a complete instrument or system, the repair technician has to determine the faulty subsystem or unit. To do that, they require the maintenance manual, the detailed circuit diagram and the necessary test equipment for that instrument. With these faultfinding aids, the technician can detect and identify the fault. It is important to define the fault accurately (see the example given above).

3.4.2 Troubleshooting components and circuits

Figure 3.4 shows the different ways by which one can troubleshoot an electronic circuit. Obviously, faultfinding technicians do not use the first

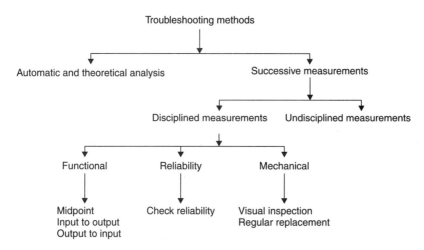

Figure 3.4 Faultfinding methods

method. The method of undisciplined measurements is not recommended, although it may sometimes be useful where we have what is called a historic fault; this is a repetitive fault in a certain type of equipment. Yet even here one should still strive to find the real cause for the fault, so it doesn't happen again.

The reliability-centered method is useful yet it is more expensive. An economic study should be carried out to justify the extra expenditure. The functional method is the most widely used. The choice between the different methods depends on the fault at hand.

Power supply circuits

The basic block diagram of a power supply unit is shown in Figure 3.5. The four main blocks are the transformer (usually a step-down transformer), the rectifier unit, the filter unit and the regulator unit. Here are three types of power supply unit:

- The basic power supply unit

- The linear stabilized power supply unit

- The switching mode power supply unit

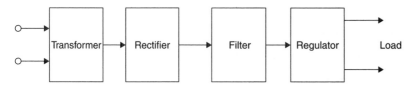

Figure 3.5 Power supply unit

When a power supply goes faulty, the fault can be in any of the four basic building blocks, yet it is useful to begin diagnosis by making some simple tests like these:

- Measure the DC output voltage; if it is zero then measure the input AC voltage.
- Check the fuse.
- If all is okay, use the half-split method to determine the fault (Chapter 8).

Amplifier circuits

The amplifier uses a small input signal to control a large output signal or power. The amplifier building blocks are the active device (transistor), the power supply and the load resistor. Many types of amplifier are used in electronic circuits; here are some of them:

- Audio frequency amplifiers
- Radio frequency amplifiers
- Video and wideband amplifiers
- Direct current amplifiers

Amplifiers may also be classified by the way they operate:

- *Class A*: the active device is biased so that current flows all the time; it is used in small-signal amplifiers.
- *Class B*: the active device is biased at cutoff so that current flows for only one half of the input signal; it is used in push-pull power output amplifiers.
- *Class C*: the active device is biased beyond cutoff so that current flows only when the input signal exceeds a relatively high value. It is used in pulse switching and transmitter circuits.

When an amplifier becomes faulty, the service technician must have the following test equipment before attempting any faultfinding work:

- Stabilized power supply
- Multirange test meter
- Square and sine wave generator
- Variable attenuator
- Oscilloscope with appropriate bandwidth

If it is needed to perform special measurements such as distortion, noise, stability, pulse response or any other test, it is important to have the following additional equipment:

- Distortion meter

- Noise measuring meter

- Spectrum analyzer

- Phase meter

- Function generator

To perform faultfinding on an amplifier, it is useful at first to monitor the effect of the fault on the performance of the amplifier. Table 3.1 gives some of the symptoms and the expected faults that generated them.

Always perform several tests to determine the fault accurately; here are some possible measurements:

- Gain

- Frequency response

- Bandwidth

- Input impedance

Table 3.1 Some symptoms and their expected faults in amplifiers

Symptom	Probable cause
Large change in the amplifier operating point Distorted signal No output	Failure of bias component Open-circuit resistor High value of resistor
Large change in the amplifier operating point that drives the active element to conduct too much Distorted signal	Decoupling capacitor short-circuited Decoupling capacitor open-circuited
Excessive gain Instability	Feedback circuit open
High hum (100 Hz)	DC power decoupling capacitor open-circuit
No signal in final stages Bias is normal	Open-circuited signal-coupling capacitor
Reduced bandwidth Poor low-frequency response	Change in the capacitor value

- Power output

- Distortion
  - amplitude
  - frequency
  - phase
  - crossover
  - intermodulation

Oscillator and time base circuits

The oscillator produces an output, which varies its amplitude with time. In fact, an oscillator is an amplifier with a positive feedback loop, a network to control the frequency, and a source of DC power. The main building blocks of an oscillator are shown in Figure 3.6.

The frequency stability of an oscillator is a very important element in its operation. Many factors can cause a frequency drift, such as:

- Change in the DC power supply level

- Change in the load

- Change in the transistor's parameters

- Change in the values of the components that control the frequency (capacitors, resistors)

The basic measurements needed to help in faultfinding are:

- Frequency measurements

- Harmonic distortion measurements

- Square and pulse waveform shapes

When an oscillator becomes faulty we have to monitor the effect of the fault on its performance. Table 3.2 gives some of the symptoms and the expected faults that generated them.

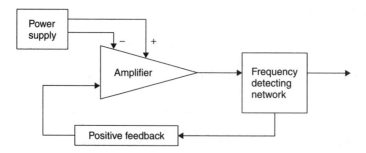

Figure 3.6   Oscillator

Table 3.2 Oscillator and time base faults

Symptom	Probable cause
Zero output Correct DC bias	Open positive feedback loop
Increased output amplitude	Open negative feedback loop
Zero output Incorrect DC bias	Open-circuited bias component
Poor frequency stability Some change in the frequency	The crystal (in the crystal oscillator circuit) is open-circuited

Pulse-forming and wave-shaping circuits

There are many pulse-forming and wave-shaping circuits; they fall into the following categories:

- Linear passive circuits (R, C, L)

- Nonlinear passive circuits (diode clippers)

- Active circuits (transistors, switches)

The faults that happen in pulse-forming and wave-shaping circuits are different than the faults discussed up to now; this is because the signal may get degraded, or it may suffer spurious triggering, which affects the resulting pulse shape without any change in the DC bias settings. The fault can also be caused by the presence of noise or interference.

Now the standard faultfinding process might contain the following steps:

1. Measure the power supply voltage and ripple content.

2. Check the input signal presence and shape.

3. Check the condition of the input leads and ensure proper earthing.

4. Check the different stages of the equipment individually.

5. Check the condition of the transistors by measuring the parameters (without desoldering).

Thyristor and triac circuits

Thyristors, or silicon-controlled rectifiers (SCRs), and triacs are four-layer semiconductor devices used in the circuit as high-speed power switches; they are used to control large amounts of power. One of the important applications of the thyristor is in the smooth control of AC power, while the triac is used in full-wave AC control circuits.

Table 3.3 Thyristor and triac faults

Symptom	Probable cause
Gate signal is high Thyristor off Thyristor can't be triggered to conduct	Gate-to-cathode open circuit
Anode-to-cathode voltage drop is zero Thyristor conducting in both directions	Anode-to-cathode short circuit
Thyristor off	Anode-to-cathode open circuit
Gate signal is zero Thyristor off Thyristor can't be triggered to conduct	Gate-to-cathode short circuit

Thyristors and triacs usually fail for thermal reasons. They can be destroyed completely when they are subjected to overload current surges. Here are the faults that can happen with thyristors and triacs:

- Low-voltage forward breakdown

- Gate control loss

- Open- or short-circuited anode to cathode

- Open- or short-circuited gate to cathode

Table 3.3 gives some of the fault symptoms and the expected faults that generated them.

Circuits using analog and digital ICs

The advantages of integrated circuits (ICs) over discrete circuits are due to the large number of active elements that can be fitted into the small space of the IC. This gives a great reduction in cost, increased reliability and the possibility of miniaturization. The IC industry began with small-scale integration (SSI), progressed to medium-scale integration (MSI), then large-scale integration (LSI) and then very large scale integration (VLSI). The integration capacity doubles every 18 months.

Analog ICs are used in analog circuits, where the input signal can vary according to the design of the circuit. Digital ICs are used in digital circuits, logic and computer systems. The inputs are either high = logic 1, or low = logic 0.

Although they are designed for very high reliability, faults do occur through failure of an element inside the IC. Handling ICs is a task for the experienced technician; an unskilled technician can damage new ICs when making a repair.

Faultfinding of IC circuits can be performed step-by-step using a logical procedure; this requires certain faultfinding aids, such as:

- IC inserter
- Test clips to reduce the risk of accidental shorts
- Logic probes to indicate the state of the circuit under test
- Logic state monitor that clips on the IC and displays its state
- Logic pulser to force a change of state at the input while monitoring the output

The IC's faultfinding process can be performed according to the following steps:

1. Measure and verify the correct value of the power supply at the IC pins.
2. Check the presence of the input at the specified IC pin.
3. Check the presence of the output at the specified IC pin.
4. Check if there is a short or open circuit in the copper track connecting the IC to the printed board.

Digital logic circuits

Faultfinding digital ICs follows the same general rules stated above, yet there are some precautions to observe, such as:

- Never use large-tip probes, they may short two adjacent IC pins.
- Carefully note the maximum and typical power supply voltage for the IC you are servicing.
- When measuring the voltage, make the measurements at the IC pins; do not use the printed board or tracks, they may be cracked or shorted.
- Never remove an IC while power is on; the same applies when inserting the new one.
- Never apply test signals when the power is off.

The actual diagnosing process is performed by sequentially operating the IC within the system and comparing the resulting outputs with the outputs when the circuit was okay.

Op-amp circuits

Operational amplifiers (op-amps) should be treated in the same way as ICs during handling and faultfinding.

Troubleshooting transistors faster

There are several ways that help the O&M engineer to troubleshoot transistor circuits faster; please refer to Appendix A.

Table 3.4 Cable failures

Symptom	Measurement	Expected cause
Cable inoperative	Impedance	Open-circuit conductor
Low voltage	Impedance	Low resistance ($<10Z_0$)
Very low voltage	Impedance	High resistance ($>10Z_0$)
Flashing	Impedance	Breakdown due to high voltage impulse
Intermittent	Impedance	Occasional breakdown

FETs and MOSFETs faster

There are several ways that help the O&M engineer to troubleshoot FET and MOSFET circuits faster; please refer to Appendix C.

Troubleshooting electrical components

Faultfinding high-voltage cables

Cables can fail for several reasons. Table 3.4 gives the symptoms of the fault and the probable cause in each case.

Faultfinding high-voltage transformers

Small transformers are made in large numbers for use in electronic equipment. Maintenance for small transformers is really minimal, yet the regular inspection of certain features is vital. Since oil forms part of the insulation material of most transformers, it should be checked regularly, as oil tends to deteriorate over time:

- *Atmospheric conditions*: poor ventilation results in condensation inside the transformer, which also promotes acidity and sludge.

- *Operating temperature*: prolonged work at high temperature leads to the development of acidity sludge; this has a disastrous effect on the solid insulation.

- *Presence of moisture*: this has the effect of reducing the electrical strength of the oil.

Measuring the insulation resistance can test insulation. Yet the insulation resistance reading might be high when the transformer is cold, dropping rapidly as the maximum operating temperature is approached. The dielectric power factor measurement is also indicative in checking the state of internal solid insulation.

Here are some tests that should be carried out on the transformers to ascertain their ability to function properly once they are purchased:

- *Impulse test*: discovers many weaknesses in the transformer, usually due to poor work or poor materials, but definitely quality not design faults.

- *Temperature test*: insures the transformer will function properly at full load and insures the temperature rise is within the specified values.

- *Noise level measurement*: insures the transformer is free from harmful internal discharge within the insulation structure.

3.4.3 Cable fires and cable thefts

Cable fires

Power cable fires can be reduced by:

- Dimensioning the cable properly to safely carry full load. To start a fire, on its own, the cable must continue to be severely overloaded for a long period.

- Using cables with improved fire characteristics, especially when the cables are running in enclosed spaces.

- Insuring the cable is not damaged. Cables are seldom the cause of fire unless they are damaged or they have an internal electrical fault.

- Insuring that environmental conditions do not change the operating conditions of the cable. For cables running in collieries, the precipitation of coal dust on the power cables reduces their heat dissipation characteristics and increases the probability of cable fires.

Cable thefts

Power and telecommunications cables are targeted by thieves, especially in developing countries. Cable thieves sell the high-purity copper of the telecommunications cables to companies, which then use it in manufacturing other domestic appliances. This is a grave problem when the telephone outside plant network is overhead. With the gradual change of the outside plant network to fiber-optic cables, the problem is decreasing.

The problem of power cable theft is being addressed by shortening the distances of the 110/220 V distribution network, and by increasing the number of step-down transformer substations so the major part of the distribution network utilizes high voltage.

3.4.4 Fire

Reducing fire incidence

The first step in eliminating the risk of fire is to reduce the probability of fire incidence. This necessitates the proper storage of chemicals and other inflammable substances, since some of the fires that happen in telecommunications installations begin in the stores or the battery room. Thus the battery room should be well ventilated, and temperature and humidity control should be installed there. It is also important to prohibit smoking on the premises.

Controlling fire propagation and spread

Fire begins with ignition from an agent. If detected by any of the organization's fire detectors, there will be an alarm in the form of a sound signal, flashing light, recorded message or any other means. Usually the firefighting equipment will be alerted automatically, but this is not always the case.

If the detectors fail, fire will spread and the danger of smoke will be imminent; this will make firefighting very hard and will also affect the occupants of the building on fire, who will suffer and may even suffocate. Yet if the manual extinguishing begins before the critical time ends, there will be a high probability the fire will be extinguished with minimum losses. If the fire is left to spread due to the inefficiency of the fire-extinguishing equipment or the delay in calling the fire brigade, then large losses (or maybe fatalities) are expected due to human injuries, in addition to loss of equipment and property. Figure 3.7 illustrates the fire spread process.

Providing adequate escape routes

The effect of smoke on the occupants of a burning building is sometimes greater than the effect of fire itself, since the occupants may suffocate from smoke inhalation before fire reaches them. Thus it is very important to provide fire escape routes. The design of an escape route is discussed in Chapter 9.

Installing a fireman's switch

A fireman's emergency switch must be provided in power line-frequency circuits, supplying exterior electrical installations, as well as interior discharge lighting installations. The switch must be installed outside the building, it must be clearly marked in a red color, and it must be installed so it prevents unintentional reclosure once the fire brigade has switched it off.

Escape route planning

This subject is discussed fully in Chapter 9.

3.4.5 Arson

Arson is an act whereby a person deliberately sets a building, some equipment or an organization on fire. Targets for arson can be any

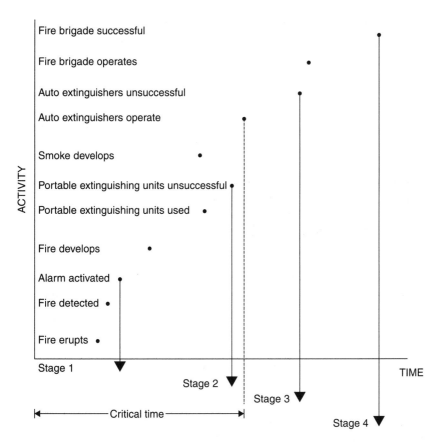

Figure 3.7 How fire spreads

organization, establishment, company or administration that the culprit would like to destroy. Warehouses and storage premises are usually the main targets for arson, since they are low-traffic areas.

Increased incidence of arson

Arson has become a significant cause for disaster in developing and developed countries. If we take only the cost of fires originating due to arson, they constitute between 15% and 50% of the total cost of all fires.

Remedial measures

The enormous increase in arson can be seen as part of the general breakdown of law and order characterizing our society nowadays. What we must look forward to, as a long-term strategy, is to make fundamental changes in the attitudes and behavior of society to achieve an ultimate cure. This necessitates the following:

- Changing people's attitudes and behavior by new policy thinking and new government programs. The best way is to begin by educating children, which requires teacher education before that. It is a long-term policy.

- Actively involving local governments, regional bodies and community action groups in the fight against arson on the grounds that arson activities are affecting their own neighborhood. Such involvement has already been initiated in the United States by the local arson task forces (LATFs) and in Europe by similar local agencies.

- Developing management action plans against arson in each establishment, organization or administration.

Motives and targets of arson

Vandalism

When young people are involved in arson it is usually due to vandalism; this is most common in schools. Take this into consideration when you have young apprentices in the organization.

Revenge

A far greater percentage of arson is caused for revenge than for vandalism.

Profit

Arson may be used to disguise a theft. But new forensic equipment makes it harder to conceal evidence of the break-in.

Mental instability

The arsonist may be mentally unstable due to the effects of drugs or alcohol, or through a mental or psychological illness.

Concealing another crime

Fire is also used to conceal another crime. If the storekeeper wants to conceal the loss of a large number of ICs in the store, they may set the store on fire to conceal the loss.

Arson control

Security equipment and systems

A complete and appropriate security package should be considered, one that contains the following items as a minimum:

- An intruder alarm system, to be connected with a central security station.

- An automatic fire detection system connected to the same central station.

- Closed-circuit TV cameras located at strategic positions.

- Appropriately illuminated potential break-in areas, with minimum dark spots; this will reduce the probability of arson, since arsonists usually target dark areas.

- Separate protection for the central security station premises and equipment.

Design of premises

A lot can be done to design the working area to make it hard, or at least difficult, for arsonists to operate easily; some of these measure are:

- Positioning vulnerable operations so they are hard to access by unauthorized personnel.

- After working hours the area should be well closed to keep out intruders.

- Careful attention must be paid to doors and windows, especially how and where locks are fitted and their quality; they shouldn't be easy to open with, say, a credit card.

- The availability of firefighting equipment at strategic locations.

- If the power supply is provided by diesel generators then fuel storage — an important factor in fire safety — will be the main target for arsonists; fuel storage should never be placed in an easily accessible position, e.g. near the main road.

- Refuse and waste are used by arsonists to initiate fires; do not allow easy access to refuse and waste.

3.4.6 The computer system

During the last half-century the world has progressed more than in any other period in recorded history. One of the reasons for this rapid advance is the information revolution, made possible through the capabilities of the computer.

Information technology (IT) is now part of every aspect of our lives. It has its advantages, which we can never deny, but it also has certain disadvantages, since it has opened the door for antisocial and criminal behavior in ways that are completely new to business and the community at large.

Unfortunately, computer systems give criminals the means to commit traditional crimes in a nontraditional way. The result will not only be the economic loss due to the computer crime, but also the suffering of the people who experience interruption. Our security systems have not kept pace with this rapid technological change.

Computer problems

The main problem with computers is that they are now a vital element in everyday life. There is hardly any activity that does not use a computer in some way or other. Thus, when the computer system fails, disaster occurs. With computers we have to differentiate between *computer misuse*, the accidental or negligent misuse of the computer system, and *computer abuse*, the intended unauthorized access to the computer system to commit something unlawful.

The second problem is computer crimes, because they are somewhat invisible:

- The very large storage capacity of the computer, and the constant striving to increase the speed of its operations, makes it very hard to detect computer crimes.

- The computer crime investigating team seldom has the necessary expertise to conduct an in-depth investigation, which should be done by professional programmers who are usually busy with software development.

- Many companies and certainly most individuals do not have contingency plans to respond to computer crimes.

- Several computer users fail to admit or even acknowledge the existence of this security problem, because:
  - They fear the associated negative publicity and public embarrassment.
  - They fear the loss of new investors or public confidence, and the consequential economic loss.

Computer criminals

Those who commit computer crimes can be students, amateurs, terrorists, but most probably they are company employees; that is why computer crime is usually an insider crime. One statistic shows that employees of the victimized companies commit 90% of the economic crimes. Another shows that internal sources contribute 70% of the computer security risk in any company; external sources account for only 30%.

Computer crimes

The introduction of computer viruses aggravated the situation since a virus makes the computer software act according to its commands, which poses real threats. Sophisticated viruses can be targeted for specific objectives at specific industries to commit a variety of traditional crimes, at a specific time or at the initiation of a certain command.

Losses and damage

A survey conducted in 1987 showed that 24% of corporations and government agencies were victims to computer-related crimes in that year, with losses ranging from US$ 145 million to US$ 370 million. Another survey conducted at virtual address extension sites (VAX) in 1991 showed that 72% of the respondents suffered a security incident within that year, and 42% indicated that the incident had a criminal nature.

Yet these statistics do not give the whole picture, because the affected organizations fear the negative publicity and public embarrassment, as well as the loss of new investors or public confidence, and the consequential economic loss.

System vulnerabilities

The economic value of data stored in a computer is now a recognized fact, thus when a computer fails due to criminal offense, the value of the replacement parts that were stolen becomes only a very small proportion of the economic loss suffered due to the loss of, or inaccessibility to stored data, due to information misappropriation or damage. There are several factors which can make a computer vulnerable to threats.

High storage density

High storage densities mean a single error can have enormous impact. With huge volumes of data, it can generate considerable losses.

Ease of accessibility

Each computer system must be accessible to different users, especially those contacting the computer from remote terminals; these are attacked by:

- Using false identification and access codes to retrieve the needed information.
- Unauthorized use of a terminal logged in by an authorized person who left it logged on and unattended.
- Maintenance personnel tampering with hardware or software to get extra privileges.

Maintenance personnel have direct access to computer terminals; they usually work after hours so they do not disturb the normal functioning of the business.

Network complexity

Computer operating systems contain millions of individual instructions, so the possibility of finding an error is high; such an error can result in system downtime or even a potential security fault. Hackers take advantage of this and use these program weaknesses to attack the system, causing major losses to organizations.

Vulnerability risk

Computer systems as well as all electronic systems are reliability centered; they are susceptible to component reliability figures, component aging, physical damage, environmental effects, electromagnetic interference, and interception. Computer data can be changed instantly or deleted with very little chance of detecting the culprit.

Human effects

Employees contribute the bulk of computer crimes that happen in their own organizations. It was discovered through research that criminals would be reluctant to commit a crime to hurt individuals, since they regard this action as immoral, but they approve and can find an appropriate rationale for doing harm to their organization.

Common types of computer crime

Manipulating the computer system

The most common computer fraud nowadays is the manipulation of intangible assets stored in data format, such as money deposited, hours of work performed, or credit card information. The ease of remote access to computer data banks has tempted criminals to try various computer-related frauds.

Input manipulation is simple and easy to perform, but program manipulation needs sophisticated computer knowledge. Program manipulation is hard to discover because it changes the existing program. Output manipulation deals with the computer output, such as cash-dispensing machines.

Forgery

Forgery changes or alters the data stored in the computer. Alternatively, the computer itself can be used as an instrument of forgery by creating false documents or banknotes.

Data and program damage

Some criminals gain direct or indirect access to an organization's computer system using special programs to modify, suppress or erase items. This unauthorized access is intended to create a disaster in the organization.

Illegal access

There are many motives for illegal access to a computer system: curiosity, sabotage, espionage, etc. Some hackers are just curious to see how far they can go. Sabotage is a criminal act. Espionage is also a criminal act; it is intended to disrupt the operations of the organization or to acquire vital information that may help competitors gain technical advances. Illegal access can be accomplished by several means:

- Through improperly protected telecommunications networks.
- Through undetected loopholes in an existing security system.
- Through impersonating legitimate system users.
- Through legitimate access points or trapdoors created for system maintenance.

Illegal access through passwords can be accomplished by discovering (or stealing) a password that allows entry to the system, then using a special program to capture the legitimate user's passwords; or by using a password-cracking program.

Preventing computer crime

Computer security has vital importance to the organization; by 'security' we mean the accidental or intentional unauthorized access to the data or information stored in the computer system, access which might threaten the system's confidentiality, integrity or availability. The aim of computer security systems is to reduce, to an acceptable level, the risk of accidental or intentional unauthorized access to the system. Computer security systems must protect:

- Information and data stored in the system as well as the software developed to perform the work, to preserve confidentiality, integrity and availability of the system.
- The data processing services, which can be considered the organization's most important asset requiring protection.
- The data processing equipment.

Develop an overall security policy that covers the following items.

Personnel security

- Specify the security requirements commensurate with a person's job duties, responsibilities and authorities.
- Screen company personnel to insure they meet the security measures.
- Supervise systems access and control.

Premises security

- Choose the site location and plan the layout to cope with the security measures.

- Identify and implement procedures to control access to the restricted areas.

- Identify, prepare and implement a disaster management plan to safeguard the organization from damage due to fire, flood, explosions, attacks and any other physical damage.

- Identify, prepare and implement a disaster management plan to safeguard the organization against failure due to power interruptions and other failures that might affect the security system's operation or efficiency.

Operational security

- Identify existing and anticipated risks.

- Prepare security contingency plans.

- Properly define the security duties and responsibilities within the plan; this includes the following features:
  — Identifying authorized users
  — Restricting each user to the area they are authorized to access
  — Regular monitoring of the system operations to detect any security violation
  — A counterattack facility that helps the system cope with harmful security violations

- Creating restricted areas for company computer systems.

- Creating and using properly engineered authorization procedures.

- Creating the necessary organization to administer and implement the security plan.

Firewall systems

A firewall is a security system that permits an access control policy to be enforced between two networks. This enforcement can be to block only unauthorized traffic, i.e. to explicitly deny all services except those critical to the mission of connecting to the network, or to permit only authorized traffic, i.e. to provide a metered and audited method of queuing access in a nonthreatening manner. The access control policy must be specified by the organization utilizing the firewall system. Thus a company using a firewall system will be secure against unauthorized interactive logins from the outside world.

The level of monitoring, redundancy and control should be specified according to the organization's objectives. The financial issue should also be discussed; cost estimates for the proposed firewall system should be discussed in terms of what it will cost now and the continuing costs such as support.

On the technical level, a decision must be taken about the traffic routing implementation. Will it be at the IP level or at an application level? Finally, there are two basic types of firewall:

- *Network-level firewalls* control access to and from a single host by means of a router operating at a network level (Figure 3.8). The single host is a highly defended and secure strong point that can resist attack.

- *Application-level firewalls* are networks of screened hosts (Figure 3.9) in which an application gateway or forwarder (proxy server) permits no traffic directly between networks, and it performs elaborate logging and auditing of any traffic passing through.

The firewall system cannot protect against the following possibilities:

- *Attacks that didn't pass through it*: once you install a firewall, all communications must be routed through that firewall to insure maximum security.

- *Traitors or idiots*: some people may use the network in a careless or negligent way, such as copying important files onto floppy disks.

- *Viruses*: the organization should use antivirus software that must be updated regularly to safeguard its computer data and information from disaster, due to any virus attack. Do not rely on virus-detecting firewalls.

Protect the system against Trojan horse viruses. When a Trojan horse virus is installed on a PC, it can observe sensitive data before it is encrypted

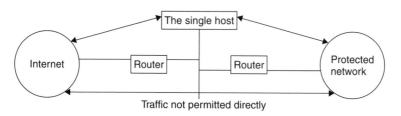

Figure 3.8 Network-level firewall

Figure 3.9 Screened-host firewall

by the virtual private network. This means that when at home you encrypt sensitive information to send to your organization, the Trojan horse virus observes the data before it is encrypted, unless your PC has a firewall. In this case the PC is used as an intermediate cloaking device.

The revelations in February 2000 that the former director of central intelligence in the United States improperly handled classified information on his home computer raised awareness about the grave consequences of this new form of hacking. Telecommunications companies are now aware that telecommuters can pose grave security risks, because hackers are using them for improper intrusion into the system. Hackers are not only accessing the corporate network illegally, but are also using a new malicious program to cover their tracks.

For this reason, and as a disaster prevention measure, many companies are now making it mandatory that each telecommuter's home PC be provided with a personal firewall program to secure home PCs accessing corporate networks, and to track invaders who seek to use them in attacks.

The year 2000 problem

The year 2000 (Y2K) problem is a computer software problem that originated in the early days of the computer era when programmers were trying to economize on the small memory capacity available at that time. Thus the databases of the computer systems that run worldwide business and communications were designed to carry only the last two digits of the year (e.g. 99 instead of 1999). By the time memory capacity increased, the programmers forgot to mend this simple error, thus the year 2000 would be represented in the computer program as 00.

To a computer, 00 is zero not 2000, so time reverses itself: earlier dates become later dates, events of last year seem to have happened 99 years ago (or won't happen for 99 years), credits become debits, and the computers go crazy.

Early symptoms of Y2K were detected in most computer systems several years ago, and computer users were advised to watch for certain warning signs to prevent serious damage to their organization. There are seven distinct signs.

Precedence/comparison sign

Using two digits for the year is okay when you want to decide which of the two years is earlier, as long as both years are in the same century (e.g. 1978 is later than 1968, because 78 is greater than 68) so a computer doing a *comparison* will get the correct answer. But although 2001 is later than 1999, **01 is less than** 99, so in comparison the later year would seem to be the earlier year. The computer sees time as having reversed itself; later time becomes earlier, and earlier time becomes later. Thus, in any logical comparison or test, the computer will make the wrong choice every time.

Sorting sign

Using two digits for the year causes no problems when you need the difference between the two years, as long as both years are in the same century (e.g. $1966 - 1964 = 2$ and $66 - 64 = 2$); the two methods produce the same result. Yet in the Y2K problem this *simple subtraction* does not work when the two dates are in different centuries (e.g. $2001 - 1999 = 2$ but $01 - 99 = -98$). Thus instead of 2, we get -98. All sorts of funny (as well as disastrous) things might happen. A two-year-old boy may become a 98-year-old grandfather.

Special numbers

Programmers usually use 00 to indicate there is no entry, or that the item under consideration has no specific due date, so in this case they write the date as 00/00/00. When the date is inapplicable or unknown they write the entry as 99/99/99. They might include a line in the program that says: If the year is equal to 99 then skip the normal processing. In this case you obviously have a problem for all items, even in the year 1999. Also, in many file storage systems, 99 is treated as a special number meaning 'never', e.g. if a file is given an expiration date of 31/12/99, this means *never* let this file expire.

The last shall sort first

On sorted lists, the items from the twentieth century and the items from the twenty-first century will trade places. In your *sorted list*, for example, the oldest items will be the items from the year 2000, even when items from 1999 are on the list because, for the computer, the 1999 items happened 99 years after the items of the year 2000. This could lead to problems ranging from minor annoyance to items being misapplied and thus incurring large losses.

Interface problems

Interface problems are usually a reflection of other problems, yet if you have a client/server system, the interface will be located on an entirely different platform, possibly in another country or continent. The fact that two-digit years will enter a valid range of days and months may introduce ambiguity to your system. The dates 97/03/17 and 03/17/97 are both clearly indicating 17 March 1997, but what about 01/02/03? Is it in 2001 or 2003?

If you mix date formats, or are international in scope, your incidence of erroneous interpretations will increase dramatically in the year 2000. Moreover, this kind of problem will start with entries from the year 2001 not 2000, so we have to watch for it.

The days of the week

Dates in the twenty-first century will lag their twentieth century equivalent by one day, since there are 36 525 days in the century (when the double-naught year is a leap year), and 36 525 divided by 7 (days of the week) leaves a remainder of 6 days. If there had been one more day in the century, we would have been fine. But due to this discrepancy we get the one-day lag, thus 1 January 2001 was a Monday, but 1 January 1901 was a Tuesday. Yet because the year 2000 is a leap year, during the period 1/1/2000 till 28/2/2000 we will have a lag of two days, thus 1 January 2000 was a Saturday but 1 January 1900 was a Monday. So if any of your two-digit year programs are determining the weekday, they are probably assuming a date in the twentieth century as the base, and will be off by one day (and by two days in the first two months of year 2000). This might affect the maintenance scheduling programs, for example.

The programmers problem

The programmers problem is the most important problem of them all, because no one knows what some programmers might have done to the date fields in some programs. The date or year is sometimes embedded in a key field used as an identifier for a record. That might be a problem if the key field is a storage key for the record on the database. In some storage systems the actual storage location is calculated from the key. This means that if you expand the year, your database records will be in a randomly determined wrong location in the database. Therefore, the records will have to be removed as well as expanded, which could increase the demand on your system exponentially.

Scope of the Y2K problem

The reports of the world's demise due to Y2K are exaggerated, as we now know. Computer glitches are nothing new; we have all encountered situations or heard stories of a computer bug that caused some problems. Yet we have to discuss the seriousness of Y2K and what we need to do in order to minimize its effects. Here are some of the factors to consider.

The magnitude of the problem

In some normal computer glitches, a few small errors may go undetected for a long time. But with Y2K it is expected that huge errors will affect a large number of records. The magnitude of the problem will lead to its immediate detection, so be ready to take quick action to neutralize the exposure, and then you can calmly figure out a way to redress the problem.

How the computer can fail

Computers can do only what they are programmed to do, and programming them is hard work. Programs may not realize a *sort* is out of order, because sometimes it is grouping not order that is important. Also, due to a date bug, a computer might not recognize that an extremely young person is getting a pension, because the programmer did not program the computer to look at the birth date before writing a pension check.

How the Y2K problem was confronted

The Y2K problem was not expected to bankrupt us, but it was sure going to cost us a lot, thus it was a good idea to overreact than to underreact. The US Congress passed a bill, signed into law in October 1998, authorizing an emergency spending of US$ 3.4 billion to fix the Y2K problem in government computer systems. It raised the budget allocated for Y2K to US$ 5.4 billion; of this, US$ 1.1 billion was earmarked for Department of Defense systems and computers supporting national security.

Thus, Y2K was expected to be a major project. The worst-case scenario is that some companies will lose a lot; the best-case scenario, if all goes just right, is zero return on the investment, i.e. no ROI. Yet the following potential disasters may be good reasons to ignore the zero ROI criteria:

- *Railroad scheduling*: a train wreck could kill many people.

- *Airline scheduling*: flight upsets could have disastrous consequences worldwide.

- *Bank wire transfers*: an erroneous transfer could deplete an account.

- *Interest and maturity*: faulty calculations could cause havoc in banks.

Some solutions for Y2K

Replace the system

Every system has a limited working life and there comes a time when you have to replace it. This might appear to be the cheapest solution, since it would be more expensive to fix it than to replace it, after all the system is probably at the end of its useful life anyway. If a replacement is due, you effectively fix the Y2K problems for free, simply by replacing the system with a new Y2K-compliant system. Here are two advantages:

- You get a whole new system, with the latest technology.
- You get some return for your Y2K expenditure.

Here are two disadvantages:

- A new system might be expensive.
- You might not have the time even if you have the funds.

Expand the date field

Expanding the date field to allow four-digit years is the second and cleanest approach. This involves making the database look the way it would have looked if things had been done right in the first place. This means establishing an algorithm (or a rule) to do the conversion. In this case you take each two-digit year and correctly attach 18, 19 or 20 to get the four-digit year. Here are two advantages:

- It is the complete, permanent and obvious solution.
- It removes all ambiguity from the dates.

Here are seven disadvantages:

- Record layouts must be changed and you need to convert your data.
- Record sizes must be increased, so you need to do some restructuring.
- You will need more storage.
- With each field adjustment you have to adjust everywhere.
- You must recompile all programs to adjust to the new record layout.
- All the changes must be done simultaneously for each upgrade unit (converting the data, restructuring the database, fixing the programs, recompiling the programs, etc.).
- This is usually the most expensive solution.

Fix the processor

We have discussed changing the data (expansion), now we look at fixing the processor. This is achieved by a process called windowing, because adding extra digits may not be possible. Yet it is possible to store years in two digits, and still do the calculations correctly, but we need more computer power in this case. The process is called the sliding window.

In this process we take advantage of a real-life situation. If I introduce someone to you indicating that they were born in the year 46, you automatically assume that the year is 1946, since it cannot be 1846 or 1746, for obvious reasons. We can program the computer to make the same kind of assumptions by establishing a window extending 50 years forwards and 50 years backwards, and it can slide based on the current date. Thus we make a window for each year to eliminate confusion:

- In 2000 the window is 1950–2049
- In 2001 the window is 1951–2050
- In 2014 the window is 1964–2063
- In 2050 the window is 2000–2099
- In 2184 the window is 2134–2233

What the computer does is to fit the two-digit year into the window to get the four-digit year. Here are four advantages:

- You don't have to change all the programs.

- You can fix the programs one at a time.

- No conversion or restructuring is required.

- It is similar to expanding, but there is no need to increase storage space.

Here are three disadvantages:

- All programs must use the same assumptions for current date and past and future window size.

- Stores and indexes must be addressed separately.

- There is a potential performance impact due to on-the-fly conversion.

Compression

In this process we expand the years to four digits, but then compress those four digits back to the same space originally used, thereby avoiding the need to restructure the database. Here are four advantages:

- It is similar to expansion, except there is no need to increase storage space.

- Conversion removes all ambiguity from the dates.

- It requires only minimal logic fixes.

- There is no need to expand date fields.

Here are six disadvantages:

- Record layout must be changed and you need to convert your data.

- With each field adjustment you have to adjust everywhere.

- You must recompile all programs to adjust to the new record layout.

- All the changes must be done simultaneously for each upgrade unit (converting the data, restructuring the database, fixing the programs, recompiling the programs, etc.)

- User access, queries and spreadsheets are problematical.

- Encoded data requires conversion whenever you work with it.

Encapsulation

The idea behind encapsulation is a time warp. If we were in 1971 our programs would run just fine and would roll on into 1972 without any problems. What if we fool our programs into thinking 1999 is 1971? This may be done by subtracting 28 from each year on our databases. Then we intercept input and output operations and adjust the years back and forth for human consumption. In some systems this could be done outside the program. Here are two advantages:

- You do not have to touch the programs.

- Encapsulation involves less testing.

Here are six disadvantages:

- You must have exits available to intercept data.

- If you miss any program modification path, you will corrupt your data.

- You have to intercept and adjust data before presentation to users and before storage.

- Performance may be affected by on-the-fly conversion.

- You must do a database conversion to subtract the warp factor.

- Encapsulation has the shortest effective range of all the solutions — only 72 years.

Abandoning

It may not be possible to fix all the known Y2K bugs in your system, then one should be realistic and abandon programs and systems that cannot be saved. Here are two advantages:

- You can save a lot of work.

- You may have no other choice.

Here are two disadvantages:

- You might discover later on that you have abandoned something critical.

- It may have a negative effect on strategy and public relations.

What really happened

Judging the Y2K fix in the new century, it seems to have been a reasonable endeavor, because it proved new processes and disciplines that had tremendous value beyond the problem itself. But by looking back at what really happened, we can identify some of the weaknesses in our approach and our disaster management procedures. Analyzing these weaknesses carefully can transform them into strengths when dealing with future disasters.

Before evaluating what happened, it might be a good idea to discuss the main points to consider when doing an evaluation:

- Secure management's approval and sponsorship for the evaluation exercise, otherwise you will go nowhere. It is estimated that the postevaluation project will take 1–5% of the cost of the Y2K project, but the benefits gained will be immeasurable.

- Choose the evaluation project leader; the most appropriate person would be the Y2K project leader accompanied by their team. They already have many ideas about the problems faced, solutions sought and benefits gained.

- Begin the preliminary evaluation while people's memories are still fresh and the documentation is readily available.

- The issues under discussion are rather broad; it might be a good idea to sectionalize your efforts, dealing with an issue at a time, then at the end you can integrate your findings. You could break the team into groups to study different aspects, but each group should follow the same basic process in its work. Here are some suggestions:
  — project management
  — disaster management
  — risk management
  — project communications
  — contingency planning
  — disaster recovery
  — dealing with service providers
  — disaster management training

- Conduct a brainstorming session to examine all aspects brought forward by the team members, sharing information about what went right and what went wrong, what was actually done and what should have been done.

The brainstorming session

1. Each team member must filter their discoveries and think of better action plans, with fixed responsibilities that can be followed up easily.

2. Combine the findings and ideas of all team members, then compare the implemented Y2K plan with what actually happened. Try to understand what was missed out of the original plan that caused the glitches during implementation. This will help you to improve your disaster planning.

3. Prepare an executive overview for the company managers; this will help to outline the proper way to proceed if a similar problem occurs. The global increase in e-business suggests another problem is likely.

The final analysis

- The major sources for the Y2K problem have been fixed, because programmers all over the world spent valuable time and money working on them. That is why no serious life-threatening disasters have been reported and there was no sign of the previously anticipated panic.

- The potential Y2K problem was not exaggerated; the problem was serious, it was detected well before it was supposed to occur, then efforts and funds were allocated to its solution. The outcome was minimal disturbance to business and people.

- The Y2K bug variables were numerous and unknown to the programmers. No one knew for sure any of the following parameters:
  — How many chips actually existed in the world
  — How many contained real-time clocks with date dependence
  — How many date-dependent chips might fail on 1/1/2000
  — How many date-dependent chips were actually compliant
  — What type of testing would definitely reveal all the bugs

- Fear was the controlling factor; many potentially faulty systems were turned off on 1/1/2000, or perhaps run manually, such as cash machines. Some systems were given lower loads, so that when a problem happened the standby and spare plants could handle the situation. Others were given higher than normal support and supervision; then if a problem happened, it could be dealt with immediately.

- Some of the Y2K problems have not yet had any visible consequences; we may have to remain vigilant.

- For several reasons, companies and organizations are reluctant to report the occurrence of Y2K problems after they have invested so much in compliance. They tend to deemphasize or ignore such incidents.

3.5 NATURAL DISASTERS

3.5.1 Lightning strikes

Lightning is a transfer of electrical charge. A thundercloud has a tripolar separation: a central negative charge with positive charges above and below. When the voltage between two oppositely charged fields exceeds atmospheric resistance (30 000 V) then lightning results. This may occur within a cloud, between clouds or between clouds and earth.

Lightning development

When there is a lightning strike from a cloud to earth, a negative current is carried downwards. Within 100 m of the ground, a return positive charge moves upwards. The downward strike is slower, so the observer perceives the strike to be from the cloud to the ground. Visible flashes average 5.5–6.0 m in diameter and 1.6 km in length.

Lightning formation

Lightning is formed when there is a very high field strength at a cloud base; air is ionized and a corona is created (Figure 3.10). A streamer goes from the cloud downwards to the earth. When it is near the earth, return leaders are created upwards from trees and buildings or any tall object. A conducting path is made from the earth to the cloud. A visible discharge occurs with a current intensity of up to 30 kA at a frequency of more than 100 MHz. The rate of rise of the discharge current is about 30 kA/µs.

Lightning characteristics

- Because of its high frequency, lightning will consider any bend in its path as a high impedance, and any capacitance as a very low impedance. This enables the discharge to be reflected in adjacent metallic objects, a phenomenon called side flashing.

- Mechanical forces are produced between metal parts (tubes) and the walls they are fixed to. If the tubes are not properly fixed, they may be forced away from the walls.

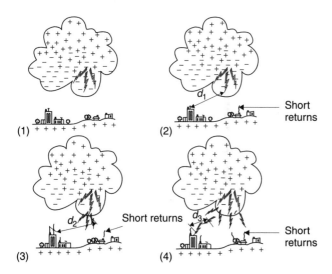

Figure 3.10 Lightning formation

- The acoustic shock wave from a lightning strike is harmful to the ears.

- Electromagnetic radiation resulting from a lightning strike is harmful to sensitive radio and electronic equipment; it may also disrupt communications.

Theories of lightning protection

Lightning research carried out in Zimbabwe, in which I participated, indicated that lightning strikes a specific part of land or certain buildings year after year, at almost the same spot. Under those lightning spots were found small pools of groundwater at depths of 3–10 m and sometimes deeper. If a pool was not found at the lightning spot, the intersection of two small water streams was found instead.

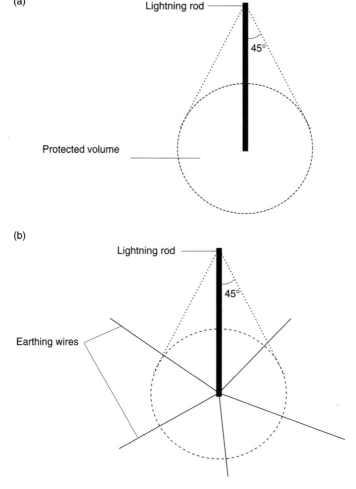

Figure 3.11 Protected volume: (a) 45° cone and (b) radial earthing wires

This discovery was very useful in planning and designing a lightning protection system for buildings. The real problem was in discovering the exact location of the pool of water, or the intersection of the two small streams underground, but without digging a hole. Dowsers proved much better than geophysicists at pinpointing the water; they found water at depths of 5–15 m with an accuracy of 15–20 cm.

A dowser is someone with supernatural powers who can identify the presence of water using two wooden sticks, one in each hand. They walk the area slowly and when the two sticks vibrate they know they are near the water spot. When the two sticks cross each other, their intersection marks the center of the pool. I have not seen a scientific explanation for dowsing, and as far as I know, it is still a mystery. All I can say is that it works.

The protected volume concept

According to this theory, a vertical lightning rod with proper earthing will protect the volume under a 45° cone centered on the rod (Figure 3.11a). But if we are protecting a very high tower (more than 50 m high) it is better to calculate the protected volume under a 25° cone.

To increase the effectiveness of the lightning protection, the lightning rod should be properly earthed. It is normal to use radial copper wires originating from the rod at ground level and radiating outwards to a distance of 20–40 m (Figure 3.11b).

The rolling sphere concept

When buildings are clustered together, we may use the rolling sphere concept to determine the protected volume. The rolling sphere concept uses a nominal protection sphere of 60 m radius (Figure 3.12).

Lightning conductors

A lightning conductor takes the energy of the lightning strike and discharges it to earth. There are five main requirements for a lightning conductor:

- Low resistance

- Straight, no loops
  — any loop will create high inductance

- Firmly secured
  — to withstand the high mechanical forces associated with lightning

- Higher than all it protects

- Diameter given by $S = I\sqrt{T/K}$ where

$$I = \text{the lightning current (A), about 100 kA}$$

$$K = \text{a constant } (K = 20\,000 \text{ for copper})$$

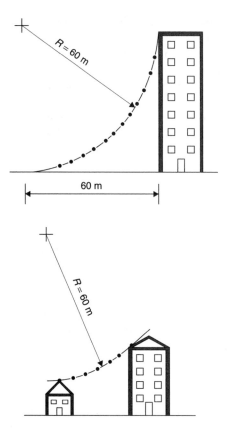

Figure 3.12 Rolling sphere

T = the strike duration (s), about 50 μs

S = diameter of the conductor (mm)

Substituting the values into the equation, we obtain

$$S = 100\,000\sqrt{\frac{50 \times 10^{-6}}{20\,000}} = 10^5\sqrt{25 \times 10^{-10}}$$

$$= 10^5 \times 5 \times 10^{-5} = 5\,\text{mm}$$

This might seem a very small diameter for discharging 100 kA, but the very short duration of the strike (50 μs) means it will not cause a large increase in the temperature of the lightning rod.

Earth electrodes

Several types of earth electrode are used to connect the lightning conductors to earth. Their advantages and disadvantages are given in Table 3.5 and some earth terminations are illustrated in Figure 3.13.

Table 3.5 Advantages and disadvantages of some lightning conductors

Type	Advantages	Disadvantages
Vertical	Easy to design and install Can be extended to reach more depth, for more stable earth resistance	High surge impedance for a single rod Is not suitable in rocky terrain Step potential on the ground is high (Figure 3.14)
Trench earth	Easy to design and install Useful when running from one building to another	High surge impedance If buried at a depth <0.5 ft will be subject to seasonal variations Can be damaged by heavy vehicles or by digging
Trench earth with radials	Has low resistance Useful when the ground is rocky and prevents vertical rods Has good radio frequency performance if laid in straight lines Has low surge impedance	If buried at a depth <0.5 ft will be subject to seasonal variations This can be rectified by terminating the radials with a vertical rod
Ring earth and foundation earth	Has low resistance Useful when the ground is rocky and prevents vertical rods Has low surge impedance and low step potential Relatively independent of seasonal variations	If buried at a depth <0.5 ft will be subject to seasonal variations Radials should be added to reduce the step potential
Horizontal grid and reinforced concrete foundation	Easy to install if done before building construction begins Has minimum surface potential gradient Has low resistance and low surge impedance	A connection from the reinforced concrete bars must be connected to the earthing downconductor, and for bonding service pipes
Utility and service pipes	If the pipes are electrically continuous, it can achieve very low resistance Has the lowest initial cost	There is no control on future alterations of the pipe connections; this might mean the earth connection is lost

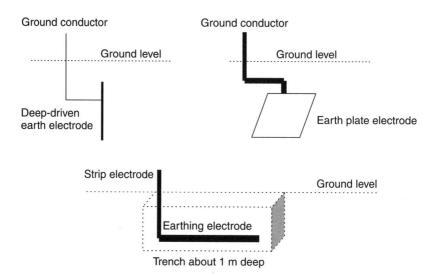

Figure 3.13 Some earth terminations

Effect of ground currents

After the strike, the energy is discharged by the ground around the lightning rod; many equipotential planes are formed there. If a person or animal passes near the rod at that moment, they will experience a current flowing through their body, since their internal resistance will be lower than the ground resistance (Figure 3.14).

Current passing through the body can produce four types of physiological damage, or trauma:

- Skin burns and necrosis of the underlying tissue; the degree of the burn depends on the energy liberated at the point of contact, and the duration of the contact:

$$\text{Liberated energy} = I^2Rt \quad \text{or} \quad \frac{E^2}{R}t$$

A high-voltage shock on very dry skin could produce a severe burn without necessarily electrocuting the victim. A lower-voltage shock on wet skin can cause death without evidence of burns.

- Paralysis of the brain's breathing center leads to deoxygenation of the blood; it is equivalent to suffocation.

- The heart may stop completely, producing a state of cardiac arrest. Failure of the heart to deliver oxygenated blood to the head will result in irreversible brain damage and then death in a very short time.

(a)

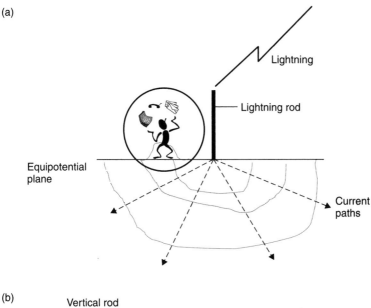

(b)

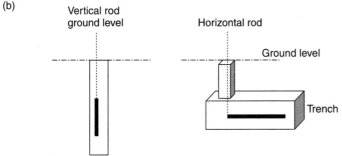

Figure 3.14 (a) Ground current, (b) earth electrodes

- When the heart tries to synchronize its beats with the frequency of the electric current, the contractions of the auricles and ventricles will lose their rhythm. A little bundle of muscle surrounding the heart ceases to contract in unison and starts to tremble or twitch. This disrupts the heart's pumping action and leads to a condition called cardiac fibrillation.

Protection against lightning

Protecting buildings

Buildings can be protected by a vertical lightning rod (Figure 3.15) or a horizontal lightning conductor (Figure 3.16). A horizontal lightning conductor should be connected to earth via an earth conductor attached to a proper low-resistance earth (less than 1 Ω). If it is not properly earthed and if its resistance is not very small, the lightning protection rod may

(a)

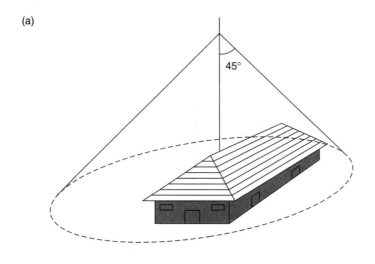

(b)

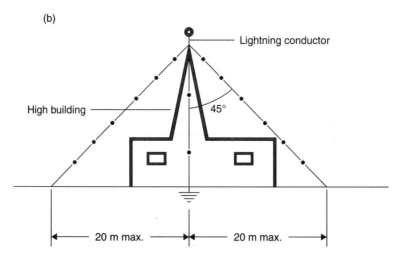

Figure 3.15 Lightning protection: (a) vertical rod, (b) tall structure

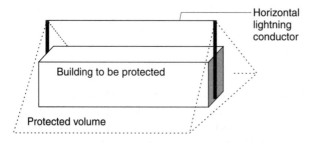

Figure 3.16 Lightning protection by a horizontal wire

(a) (b)

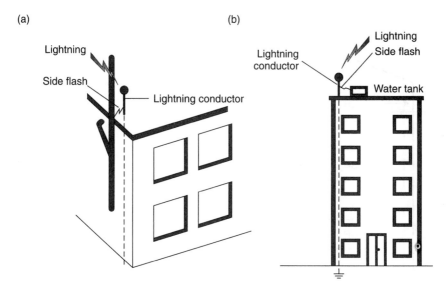

Figure 3.17 Side flash: (a) to pipe, (b) to water tank

produce some side flashes between nearby metallic objects such as water pipes and water tanks (Figure 3.17).

Protecting equipment

Each utility in the building is connected to a surge protection (SP) unit. All earth terminals are connected to a single earth to reduce circulating currents (Figure 3.18). Electronic equipment can also be protected from lightning strikes by using a box that short-circuits (SC) all input and output terminals together during a strike, and connects them to earth (Figure 3.19a). Another method is to use a box which creates an open circuit (OS) in the input and output terminals during a surge, directing the surge current to earth (Figure 3.19b). A combination solution is shown in Figure 3.19c.

Several surge protection devices are available:

- Current-sensitive devices (fuses, magnetic circuit breakers) are very slow.

- Voltage-sensitive devices (selenium suppressors, varistors, metal oxide varistors, Zener diodes) tolerate a high current rating, have fast switching times and automatic recovery.

- Voltage-sensitive short-circuit devices include spark gaps and discharge tubes.

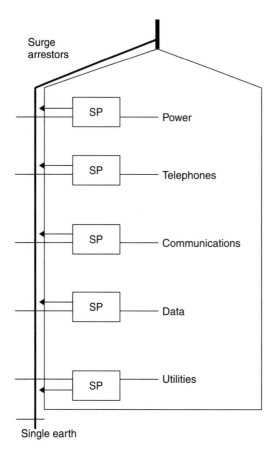

Figure 3.18 Protecting building facilities

Lightning protection drills

Besides the activities mentioned in the checklist below, the most effective drill is to let the participants watch the formation of the charge and the discharge. This can only be done in a high-voltage lab, but it does reveal three important things about lightning:

- It is a dangerous natural phenomenon but you will be unharmed if you are at a safe distance.

- The higher the voltage difference, the longer the arc. By looking at the lengths of actual lightning strikes, people can get a better understanding of the dangers.

- The sound of the discharge, when heard repeatedly in the lab, will reduce people's sense of panic when they hear the actual sound.

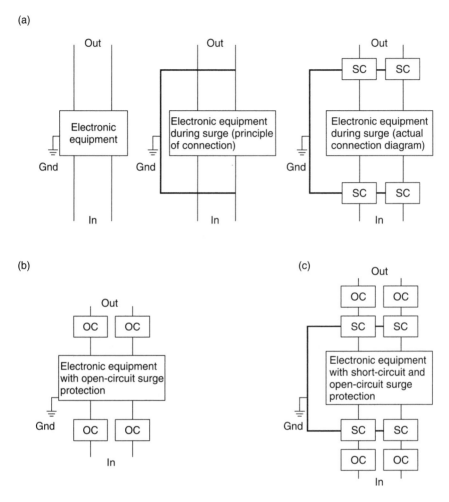

Figure 3.19 Surge protection: (a) short-circuit, (b) open-circuit, (c) combo

Checklist for lightning protection

Action before the thunderstorm

- Find out if the company is located in an area subject to thunderstorms and lightning strikes.

- Ascertain from the structural engineers that the building can withstand a moderate lightning strike. If not, examine ways to increase the building's capability to handle lightning strikes (if this is practicable and economically feasible).

- Examine ways that can reduce the effect of the lightning strikes on the company by performing certain tasks:

— Check the company earthing conductor and lightning conductor. Insure they are properly grounded. It is better to measure their resistance to ground.

— Securely fasten electronic equipment to earthing conductors; avoid creating loops.

— Hold lightning strike drills on how to avoid injury caused by unnecessary panic.

— Prepare a lightning disaster management plan, and train the employees on their tasks within the plan.

Action during the thunderstorm

• **The safest place during a thunderstorm is inside.** Since lightning usually passes along the surface of conductors rather than through them, people in buildings and cars are rarely injured by lightning. In buildings, do not stand between an open door and an open window, since lightning may travel between them. Avoid telephones during thunderstorms.

• If caught outside during a thunderstorm, **do not seek shelter under any structures that present the shortest path for lightning** (i.e. any isolated tall objects such as a tree or a tent). These serve as lightning rods, and people sheltering under them are at higher risk for both direct strikes and splash strikes. If the vicinity of trees cannot be avoided, keep clear of the tree by at least 3 m. While facing the tree, keep the feet and knees as close as possible to avoid the buildup of a step voltage across the legs. Statistics show that people get killed more frequently under a tree than out in the open.

• **If caught in an open area, get away from isolated machines,** such as tractors and diesel engines. Discard any metallic objects such as bicycles, or wet objects such as umbrellas. Get away from bodies of water. Head for dense woods or lie on the ground in a ditch, with a rubber poncho or raincoat underneath to decrease electrical grounding. You may also kneel down and bend forward with hands on knees; this presents as small an area as possible for lightning to strike.

Action after the thunderstorm

Lightning kills and injures many people each year; most lightning fatalities occur during the rainy season. Important differences exist between lightning victims and patients injured by fire or other forms of electricity. Death occurs in 20–30% of lightning victims, and is usually due to cardiac or respiratory arrest. At a lightning scene, living victims will almost always survive, but those who appear dead (in arrest) need assistance. Resuscitation attempts have a higher success rate in lightning victims than in persons with other cardiopulmonary arrests, even among patients with conventional signs of brain death. An overall resuscitation success rate of 50% has been reported.

Management of lightning victims

Delay treatment of the moving, moaning victims and instead resuscitate the apparently dead victims. The goal is to oxygenate the heart and brain until the heart regains its electrical potential and respiratory depression passes. Resuscitation efforts should continue longer in lightning victims than in victims of other trauma.

Preventing lightning injuries

Lightning injuries can be prevented by taking the appropriate precautions, as indicated earlier.

3.5.2 Earthquakes

Earthquakes move the earth; this in itself does not cause injury, but the collapse of buildings and other structures causes vast devastation. Most casualties happen due to falling objects and debris. Fire can result from fallen power lines, broken gas pipes, flammable spillages, etc. Earthquakes have a marked effect on humans due to the shock they encounter during the earthquake period. Humans can also be injured if they panic and behave irrationally.

Responsibilities before an earthquake

- Find out if the company is located in an area subject to earthquakes.

- Ascertain from the structural engineers that the building can withstand a moderate earthquake. If not, examine ways to reinforce it to cope with the earthquake (if this is practicable and economically feasible).

- Examine ways that can reduce the effect of the earthquake on the company by performing certain tasks:
  — Check the company premises and take precautionary measures by bolting down equipment that is not fastened properly to the ground.
  — Securely fasten shelves to the walls.
  — Place large and heavy objects on lower shelves.
  — Hold earthquake drills on how to avoid injury caused by unnecessary panic.

- Prepare an earthquake disaster management plan, and train the employees on their tasks within the plan.

Responsibilities during an earthquake

Earthquakes last only few seconds, yet there are minor earthquakes (aftershocks) that follow the main strike. All company employees must be prepared for them. Here are some points to consider during an earthquake:

- Stay calm since acting in a state of panic may well cause severe injury.

- Inside the company premises watch out for falling objects such as plaster, bricks, manuals and over-the-shelf equipment. Cover your head with something solid, or else use your hands.

- Do not leave the building, as you might get injured by falling objects.

- Try to put out any small fire as quickly as possible to prevent it from spreading.

- If you are outside the building, do not come near any building, wall, power poles or any other object that may fall on you.

Responsibilities after an earthquake

- Examine the casualties and losses to life or property that happened during the earthquake. Document all casualties and losses for the insurance company.

- Do not try to move any seriously injured persons since this might endanger their life.

- Check if there are any fires and put them out.

- Check the gas pipes for leaks; if there are any leaks, do not attempt to switch on electricity because the resulting spark could ignite the leaking gas.

- Check the building for cracks or damage that might threaten life. If there are long vertical cracks in the walls, evacuate the building immediately because it may collapse.

- Listen to the radio or TV to get the latest emergency bulletin.

- Investigate the scope of damage to the equipment and begin service recovery activities.

- Review the earthquake management plan in light of any feedback received during the earthquake. Amend the plan accordingly, and train the employees on their new tasks within the amended plan.

3.5.3   Tornadoes

Responsibilities before a tornado

- Find out if the company is located in an area subject to tornadoes (damaging winds that may devastate large areas).

- Ascertain from the structural engineers that the building can withstand moderate tornadoes. If not, examine ways to reinforce it to cope with tornadoes (if this is practicable).

- Examine ways that can reduce the effect of the tornado on the company:
  — Check the company premises and take precautionary measures by bolting down equipment that is not fastened properly to the ground.
  — Securely fasten shelves to the walls.
  — Place large and heavy objects on lower shelves.
  — Hold tornado drills on how to avoid injury caused by unnecessary panic.
  — Buy tornado insurance to protect the company from losses due to the tornado.

- Prepare a tornado management plan, and train the employees on their tasks within the plan.

Responsibilities on receiving a tornado warning

- Listen to the radio or TV to get the latest bulletin about the direction and strength of the tornado.

- Store emergency supplies.

- Secure outdoor objects that can be blown away by the tornado.

- Protect windows and doors so they can withstand wind-driven debris.

- Make sure the emergency power supply is ready.

- Inform end users about the likelihood of any service interruption.

- Take the tornado management plan and go over its tasks with the employees.

Responsibilities during a tornado

Tornadoes last longer than earthquakes, but their effect is comparable. The responsibilities during a tornado are almost the same as the responsibilities during an earthquake.

Responsibilities after a tornado

The responsibilities after a tornado are almost the same as the responsibilities after an earthquake.

3.5.4 Floods

The intrusion of water into the working area is one of the most common faults that can happen to a telecommunications system. The effect of water can be very harmful since all the telephone distribution cables are running under the earth. Although cable joints are usually made very professionally, there are instances when water can penetrate through the cable and damage its transmission characteristics.

Responsibilities before a flood

- Find out if the company is located in an area subject to flood.

- Ascertain from the structural engineers that the building can withstand a flood. If not, examine ways to reinforce it to cope with the flood.

- Examine ways that can reduce the effect of flood on the company:
  — Empty the ground floor (if possible).
  — Relocate critical equipment to a higher floor, above projected flood levels.
  — Identify sources for emergency service provision.
  — Buy flood insurance to protect the company from any loss due to flood.
  — Have 72-hour emergency supplies ready.
  — Stay tuned to emergency radio stations for instructions.

- Prepare a flood management plan, and train the employees on their tasks within the plan.

Responsibilities on receiving a flood warning

- Store emergency supplies.

- Move critical items from the ground floor to the upper floor.

- Make sure the emergency power supply is ready.

- Inform end users about the likelihood of any service interruption.

- Take the flood management plan and go over its tasks with the employees.

Responsibilities during a flood

- Do not attempt to leave the company since it is well equipped to handle a flood.

- Insure that vital services are running smoothly according to the flood management plan.

- Listen to the radio and TV for information about the development of the flood.

Responsibilities after a flood

- Examine the casualties and losses to life or property that happened due to the flood.

- Document the casualties and losses for presentation to the management and the insurance companies.

- Examine the state of the system and the network.

- Examine the electrical power supply condition. Do not handle live electrical equipment in wet areas. If some equipment is wet, it must be dried and checked before it is reinserted in the circuit.

- Listen to the radio and TV for instructions about help and recovery services.

- Review the flood management plan in light of any feedback received during the flood. Amend the plan accordingly, and train the employees on their new tasks within the amended plan.

SUMMARY

It is not easy to give clear definitions of major and minor breakdowns, since they depend on the types of service presented to the customer. Three important financial statements are the balance sheet, the profit and loss statement, and the cash flow statement.

The basic principles of faultfinding should be applied to electronic components such as transistors, triacs, thyristors and integrated circuits. Use logical approaches to diagnose amplifiers and oscillators.

Fire is the major cause of disaster and may sometimes be arson. Firewalls and antivirus programs help to control computer crime. The Y2K problem is a very important case of disaster management.

Follow the checklists for handling natural disasters such as lightning, earthquakes, tornadoes and floods.

REVIEW QUESTIONS

1. Define the following terms: balance sheet, profit and loss statement, cash flow statement.

2. Explain how disaster can affect the financial health of the organization.

3. Explain one way in which maintenance can be performed; indicate its advantages and disadvantages.

4. Explain one way in which faultfinding can be performed; indicate its advantages and disadvantages.

5. What are the dangers of fire and how can they be confronted? Suggest a system that will be useful at your workplace.

6. What steps should be taken to prevent the effects of arson? Suggest a system to fight arson at your workplace.

7. What are the dangers of computer crimes and how can they be confronted? Suggest a system to use at your workplace.

8. How can you prepare your organization to survive one of the disasters caused by natural phenomena (lightning, earthquake, tornado, flood)? Take the disaster you have chosen and suggest a system to tackle it at your workplace.

4

Basic Principles of Disaster Management

OBJECTIVES

- Answer questions at the end of the chapter.
- Comprehend the basic principles of disaster management.
- Identify a disaster's symptoms and causes.
- Get acquainted with types of disaster and the stages in their development.
- Discuss how to begin analyzing disaster management vulnerabilities and capabilities.
- Give examples of disasters that could happen at your workplace.
- Describe the disasters that could happen at your workplace.
- Give examples of how you would reduce the vulnerability to disaster at your workplace

4.1 ANALYZING THE DISASTER

4.1.1 The occurrence of a disaster

The occurrence of a disaster indicates a technical, administrative, financial or security failure on the part of the decision maker. Technical and human-related disasters occur for many reasons.

Random management
- *Ignoring the importance of planning*: this concept not only causes disasters but is also the main reason for problems leading to the destruction of the technical, financial or administrative capabilities of

the organization. It greatly reduces the organization's capabilities to cope with and confront disasters effectively.

- *Ignoring the organizational structure*: when the manager uses their authority improperly, bypassing subordinates, they are creating a dangerous environment in the organization, encouraging the employees to ignore their supervisors, as well as the rules and regulations of the organization, creating an excellent environment for disaster.

- *Lack of data and specific work orders*: this means that each employee will create their own set of rules and regulations for doing the work, opening a gap between the management and employees, reducing the management effectiveness in predicting the occurrence of a disaster.

- *Lack of follow-up activities*: if management is incapable of following up what is happening to monitor the actual progress and compare it with the plan, there is a risk that company objectives will not materialize. Moreover, there is a risk that corrective action cannot be taken in time to prevent the occurrence of a disaster.

Misjudgments

Misjudgments on the part of the manager, professional or graduate engineer can be very dangerous; they can be attributed to these reasons:

- *Incompetence*: when the decision maker's capabilities are not enough to allow them to take the proper decision, and they do not want to show their inability, they tend to take irrational decisions and enforce them through their authority, leading to disastrous situations.

- *Overconfidence*: here the decision maker is confident that they know best. They also do not want to involve anybody else in taking this decision, leading to disastrous situations.

- *Misjudging capabilities*: when a manager fails to appreciate the capabilities of their staff, they tend to underutilize them; they also lose their support when things go wrong, leading to disastrous situations.

Misunderstandings

Misunderstandings are major causes of disaster; they happen for two main reasons:

- Taking decisions with incomplete information
- Taking decisions hastily

Always confirm the validity of the available information to identify what sort it is. Are there postulations, opinions, viewpoints, inferences, deductions or facts? Even facts themselves are not absolute.

Misconceptions

By not fully understanding the real content of the available information and its relevance, the manager tends to take decisions with incomplete or improper understanding, in a high-risk situation.

Blackmail

Blackmailers tend to keep information to themselves then they use it to coerce people at a later date. It is an unpleasant way to treat others and it makes it hard to investigate disasters then plan for the future.

Despair

Despair is one of the psychological and sociological disasters that affect the decision maker. Despair is thus one of the main causes of disaster. When decision makers are depressed, they lose the desire and initiative to work or develop new ideas; they perform routinely without any creativity. When a disaster is in the development stage, the manager who has lost initiative fails to detect the early warning signs and leaves it to develop into a full-blown crisis.

Rumors

Rumors are one of the important sources of disasters. Perhaps there is a rumor that management has decided to reduce staff wages by 30% due to bad operating profits; this rumor could eventually lead to disaster. If the rumor spreads days before announcing the operating profits, it seems logical to the employees that the rumor is insider information about what is going on, and thus uncontrolled actions may be taken by the employees to safeguard their interests.

Power displays

Authority and power displays are two main causes of disaster. They stem from having no specific organization charts that put in detail the authorities and responsibilities of each department, section or unit in the company. Yet even when these charts exist, one of the dominating department heads may acquire some of the authorities of other department heads who do not use theirs. Thus it is the responsibility of the chief executive officer (CEO) to insure that authority and responsibility are carried out according to the company's organization chart.

Human errors

Human errors can be made mistakenly or knowingly. Training programs can help in reducing the incidence of human errors due to ignorance, but they cannot do anything about the other kinds of error. This is dealt with on pages 113–116.

Planned disasters

The concept of management by disaster, or planned disasters, is used effectively and efficiently by some managers and business people to achieve their own objectives. For example, when I was working in the Gulf countries as director of telecommunications, a wealthy real-estate developer came to the department requesting telephone service for a new residential compound of some 100 flats and several shops that he planned to build at a remote area in the desert. He was told he had to obtain a building permit from the central planning committee before his request could be discussed. When he approached the planing committee, they advised him to begin by planning the road leading to the site, and to contact the electricity and water supply authority to bring these facilities to the site at his own expense.

The real-estate developer ignored all that and began building the compound. When it was complete, several articles appeared in the local newspapers claiming that the government did not want to solve the chronic housing problem in the country since it did not want to connect the newly built compound to the water and electricity supplies. The campaign succeeded and water, electricity and telephone services were connected. The connection costs were borne by the government.

Conflicting goals

When there is no clear objective for the organization, or when the responsibilities and authorities of the top management are not well defined, many problems occur due to conflicting interests between the top management and between those responsible for implementing the decisions taken:

- *Decision makers* are the people who prepare, compile and analyze the different information and data, then they extract the indicators needed to take the decision. Pressure groups (lobby groups) do similar work so they can influence the decision maker.

- *Decision takers* actually take the decision. They are responsible to their shareholders for the validity and suitability of their decision, although they were not fully responsible for arriving at the decision, since the decision makers carried out the preliminary work. The decision taken is largely dependent on the decision taker's personality, as we will see later on.

- *Decision implementers* are the executives responsible for implementing the decision; they have no say in the decision but they implement it in the proper way. The implementation process depends on the personality of the decision implementers. They may be enthusiastic or they may be unenthusiastic; this will almost certainly affect the outcome of the implementation.

- *Decision beneficiaries* are the majority who will be affected by the decision; they constitute the environment and framework of the decision, whether they stand to gain or to lose.

Conflicting interests

Conflicting interests are a major cause of disasters, since each company looks for its interests, to maximize its profits, and in doing so one company may cause a disaster to another one. Companies with the same interests will gather together and initiate a pressure group to influence a certain decision, which they think is in their best interests. Money is spent, effort is exerted and the ultimate cost is borne by the consumer.

4.1.2 The symptoms of a disaster

Early warning signs often suggest there is a high probability of a disaster. If these signs are ignored, the disaster may be more dangerous when it finally occurs. Thus we have to monitor very carefully these early warning signs and interpret their meaning properly so that quick action can be taken to confront and solve any problems. Early warning signs, or symptoms, can be divided into several categories.

Technical symptoms

Performance testing

Before a product is produced or a telecommunication service offered, certain performance testing measurements must be made. If they indicate that the product or service conforms to the preset specifications then all is well and the product is offered or the service rendered to the customers.

When the tests indicate a divergence (even a slight difference) between the measured values and the design values, there is high probability that something wrong will happen. The main point is that any divergence should fall within the tolerance limits set by the designer; if it falls outside it, something must be done immediately.

Maintenance

The designer stipulates what sort of maintenance should be done, when, how and by whom. Failure to adhere to these instructions means that in the short (or long) term the system will fail.

Quality assurance

If the quality-of-service figures become lower than the standards to be maintained by the telecommunications authority, the subscribers will react unfavorably; they may transfer their custom to another company,

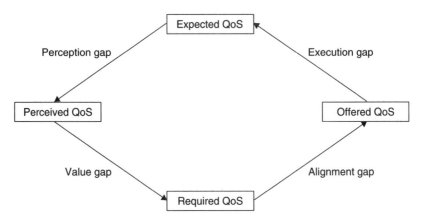

Figure 4.1   The four faces of service quality

especially when they are making a lot of international calls. This can be considered a disaster since it will affect the final profit and loss account.

The telecommunications company must therefore adhere to the quality-of-service (QoS) standards and continuously check its figures. Figure 4.1 illustrates the four faces of QoS.

In any telecommunications organization, QoS should be improving all the time, because after identifying the QoS required by the customers, the telecommunications organization offers what it considers an affordable QoS. Here we have an alignment gap between what is required and what can be offered.

The customers will judge the offered QoS against what they expected the QoS to be. Here we have an execution gap on the part of the telecommunications organization and a perception gap from the viewpoint of customers.

The customers accordingly perceive a new QoS standard, since the more the telecommunications organization offers to meet customer expectations, the more customer expectations increase. The telecommunications organization identifies a new required QoS, but here again we have a value gap between what is perceived and what is really required.

The QoS in a telecommunications company can be divided into four main parts: outside plant, switching, transmission and energy. Here are the main parameters for each part. The forms in Appendix D can be used for the calculations.

Outside plant

- *Monthly faults rate MFR*: this can be calculated (for each exchange) by taking the total number of faults of all sorts that happened during the month, and dividing it by the number of subscribers at that exchange.

- *Monthly fault-clearing rate MFC2*: this can be calculated by taking the number of faults that were cleared during the month, the same day or the day after, and dividing it by the total number of faults that were reported during that month plus the number of faults that were not cleared from the previous month.

- *Monthly fault-clearing rate MFC5*: this can be calculated by taking the number of faults that were cleared during the month within 5 days after the fault was reported, and dividing it by the total number of faults that were reported during that month plus the number of faults that were not cleared from the previous month.

- *Monthly fault-clearing rate MFC15*: this can be calculated by taking the number of faults that were cleared during the month within 15 days after the fault was reported, and dividing it by the total number of faults that were reported during that month plus the number of faults that were not cleared from the previous month.

- *Monthly fault-clearing rate MFC30 (for cables)*: this can be calculated by taking the number of faults that were cleared during the month within 30 days after the fault was reported, and dividing it by the total number of faults that were reported during that month plus the number of faults that were not cleared from the previous month.

- *Average duration of a fault ADF*: this can be calculated by taking the total duration of all the faults that were cleared during the month, and dividing it by the number of faults cleared during that month.

- *Global outside-plant quality-of-service index GOPQoSI*: this can be calculated by dividing each of the above six QoS indices by its objective, adding the results together, then dividing by 6.

Switching

- *Average period for waiting for a tone APWT*: this can be calculated by taking the sum of the waiting times to wait for a tone to dial a number, and dividing it by the number of trials.

- *Average local call completion efficiency during the charging hour ALCCECH*: this can be calculated by taking the number of successful calls made, and dividing it by the total number of trials during an hour.

- *Average local call completion efficiency during the busy hour ALCCEBH*: this can be calculated by taking the number of successful calls made, and dividing it by the total number of trials during the busy hour.

- *Average trunk call completion efficiency during the charging hour ATCCECH*: this can be calculated by taking the number of successful calls made, and dividing it by the total number of trials during an hour.

- *Average trunk call completion efficiency during the busy hour ATCCEBH*: this can be calculated by taking the number of successful calls made, and dividing it by the total number of trials.

- *Average call waiting time for operators ACWTO*: this can be calculated by taking the total time that elapsed after the dialing was completed until the operator replied, and dividing it by the number of trials made.

- *Average waiting time for a call AWTFC*: this can be calculated by taking the total time that elapsed after the dialing was completed until the call was established, and dividing it by the number of trials made.

- *Percentage of nonsuccessful calls PNSC*: this can be calculated by taking the number of communications that were not successful during a certain period, and dividing it by the total number of trials made during the same period.

- *Global switching quality-of-service index GSQoSI*: this can be calculated by dividing each of the first five QoS indices by its objective, adding the results together, then dividing by 5.

Transmission

- *Mean time between failures MTBF*: this can be calculated by taking the periods of proper functioning in hours during the month, and dividing by the number of failures during the month.

- *Mean time to repair MTTR*: this can be calculated by taking the total periods of failures in hours during the month, and dividing by the number of failures during the month.

- *Availability A*: this can be calculated by taking the period of time the equipment was functioning well during the month and subtracting the period of time the equipment was not functioning well during the month, then dividing the result by the period of time it was functioning well.

- *Global transmission quality-of-service index GTQoSI*: this can be calculated by dividing each of the above QoS indices by its objective, adding the results together, then dividing by 3.

Energy

- *Mean time between failures MTBF*: this can be calculated by taking the periods of proper functioning of the power equipment (in hours) during the month, and dividing by the number of failures during the month.

- *Mean time to repair MTTR*: this can be calculated by taking the total periods of failure (in hours) during the month, and dividing by the number of failures during the month.

- *Availability A*: this can be calculated by taking the period of time the equipment was functioning well during the month and subtracting the period of time the equipment was not functioning well during the month, then dividing the result by the period of time it was functioning well.

- *Global power quality-of-service index GTQoSI*: this can be calculated by dividing each of the above QoS indices by its objective, adding the results together, then dividing by 3.

FMEA investigations

Failure mode and effect analysis (FMEA) is a very powerful tool that helps enumerate the possible modes by which a component may fail; this in turn is used to trace the different modes with which unit failure may occur, and to identify the characteristics and consequences of each failure on the system as a whole. Chapter 5 contains more information.

Operational procedures

The operational procedures must be very clear, so they cannot be misinterpreted. Failing to comply with operational procedures can result in a disaster; Chernobyl power station in the former USSR is a classic example. Thirty-one people died instantly and thousands died later on; the whole world was endangered because some people ignored the operational procedures.

If a certain need arises that necessitates a change in the operational procedures, a special form should be completed so the change can be discussed, checked, tried, improved, approved and then implemented. The AT&T software flaw in January 1990 is a good example of changing the operational procedure before thoroughly checking the new one.

The proposition for a change in the operational procedures can take the form shown in Table 4.1. The proposition for a modification to the operational procedures can take the form shown in Table 4.2. Failing to comply with operational procedures can result in a disaster.

Contingency planning

The future, as we all know, can be very unpredictable, Murphy's law states: If things can go wrong, they will go wrong. Thus we must plan to avoid the worst that can happen to the telecommunications facility. Yet Mitchell's law, named after Martha Mitchell, states: Life is slippery like a piece of soap; if you think you have a grip on it, you're wrong.

Murphy's law indicates that things can go wrong and will go wrong, whereas Mitchell's law indicates that we have very little control over

Table 4.1 Proposition for a change in the operational procedures

Responsible manager		Date	Proposal title	Proposal reference number		
Name	Signature					
Authorization for the following assignments			By (name)	Sign	Date began	Date ended
(1) Schematic						
(2) Detailed plan						
(3) Alternative plan						
Description of the proposed alteration						
(1) Reasons for the change						
(2) Objective of the proposed change						
(3) Main points to be considered to make the change						
(4) Expected results of the proposed change						
(5) Anticipated constraints						
(6) Needed resources						

(7) Staffing (persons/day)	(8) Budget ($)	(9) Commencement date		(10) Completion date

Decision	Approved	To be revised	Implementation deferred	Decision deferred
Date	Date	Date	Date	Date

Table 4.2　Proposition for a modification to the operational procedures

Responsible manager		Date	Department/section
Name	Signature		
Title of the modification		Reference number	
Present situation			
Details of the proposal for modification			
Anticipated consequences of the modification			
Pros and cons			
Economic viability			
Decision		Date	Comments
Received			
Accepted			
Decision deferred			
Implementation deferred			
Modification rejected			
Other modifications needed			

things. Yet this does not mean we do not plan for the future. On the contrary, we have to prepare contingency plans while keeping in mind that they might not be ideal. Nevertheless, a well-thought-out contingency plan will have great benefits.

A contingency plan should address the worst-case scenario, identifying the most negative contingent events, and helping to isolate the most important factor or factors that might endanger the telecommunications system.

The first task in contingency planning is to identify trigger points which indicate when a contingent event has developed sufficient impact to allow the implementation of the contingency plan, but before a disaster erupts.

Follow these guidelines when preparing a contingency plan:

- The plan should be simple; this will make its preparation easier.

- Consider positive as well as negative contingent events, but concentrate on the negative events when preparing the plan for the first time.

- Consider the necessary actions to cope with the negative (and at a later stage the positive) contingent event.

- Consider the available resources that can help in implementing the contingency plan.

- Estimate realistically the extra resources and funds needed for implementing the contingency plan, indicating the threats facing the network if the occurrence of this negative event is not handled properly. This will help in convincing the management to approve the necessary funding.

- Recheck the plan and the implementation strategy; this may reveal unexpected vulnerabilities in your strategy, which might necessitate some change.

- Put the plan in writing to insure it is difficult to misinterpret.

- Train your staff on how to detect the occurrence of the triggering event and on how to implement the plan properly.

Training

When a disaster management plan is approved, and put into action, the company's employees must be trained to follow the steps of this plan. If this is not done, the success of the plan will be very doubtful, and there will often be grave consequences.

Training programs developed to cater for a disaster management plan should follow certain steps:

1. Verify that the problem to be addressed can be solved by training.

2. Analyze the job for which a training program is needed.

3. Analyze the population to be trained for that job.

4. Determine the entry requirements for this training program.

5. Identify the training needs and the job aids needed for training.

6. Determine the objectives of the proposed training program.

7. Design the competence tests to be completed by trainees at the end of the training program.

8. Validate the objectives and the competence tests. This is usually done on a small group selected from the target group.

9. Design the training material for the training program.

10. Produce the training material and put it through testing and development.

11. Validate the training material after evaluating any feedback from the target group.

12. Implement the training program.

13. Conduct a posttraining evaluation after the actual training program has been completed.

14. Take remedial action to rectify any serious drawbacks in the training program.

Behavioral symptoms

Loss of cooperation

Loss of cooperation among the employees is a very good catalyst for disaster. But loss of cooperation between the group responsible to plan for disaster management is the worst scenario one can think of.

Loss of zeal

When employees lose their zeal they do not innovate, they just perform their job in the way they were trained to do. Yet when a disaster is imminent they fail to detect the warning signs and they unwittingly help in its development.

Lack of responsible attitude

When people face a disaster without being fully trained to handle it, they sometimes adopt very strange attitudes. I remember an incident from 30 years ago, when a high-power transmitting station suddenly went off the air because of a faulty output tube. The responsible engineer asked the shift technician to get a new tube from the store, which he did. To save time the spare tubes were left with a reduced filament voltage, so that on switchover the tube would not take time to heat up. When the engineer asked the technician to change the faulty tube in the transmitter, with the new filament-heated tube, the technician declined because he saw that all the tubes were glowing (by filament voltage) and thought that the power was on. So the engineer had to make the switchover to put the station back on air.

Widespread faultfinding

A widespread faultfinding attitude within the maintenance group team means everyone is trying to pinpoint other people's minor faults, even the simplest oversights or the most understandable slips. Perhaps a fault situation has been set up to frame someone.

Continuous complaints

Continuous complaints usually arise from a restless O&M team. Team members complain about almost everything because they are hiding the real cause of discontent. Here the manager has to be patient with the team; they should try to discover the main reason and help to find an acceptable solution.

Spread of fear

Every time a fault occurs, somebody is severely punished; they may not be responsible but the organization must have someone to blame. No effort is made to rectify the true cause of the disaster. In this case the maintenance team will work under stress and will not devote all their capabilities to the faultfinding process; they will become preoccupied with the consequences of failure. To make things worse, the team may eventually believe they will be punished anyhow.

Increased absenteeism

Increased absenteeism can be interpreted in several ways:

- The operation and maintenance team members are not willing to go to work because they are demoralized.

- They are busy searching for another job, so they have to be absent from work every now and then.

- They feel insecure at work, so their psychological situation worsens and they feel strange symptoms of biological and psychological illness.

- They are afraid to face their supervisor and worry about the punishment that will undoubtedly follow from the management.

Feeling that all is bad

A widespread feeling that all is bad indicates a clear case of depression. Team members are depressed, so they see only the dark side of the road; they ignore the achievements in other areas of the company and look only to the disasters they are facing.

Increased customer complaints

When the maintenance team members are unhappy and demoralized, this reflects on how they treat the customers, and on the grade of service the telecommunications company offers to its subscribers. An increase in customer complaints means the service is not provided in the best professional way possible, which means the team is not performing as expected.

Organizational symptoms

Increased conflict at work

The team we are working with is usually a highly communicative group of people, with different skills and abilities. They have a common purpose and they are working together to achieve clearly identified goals.

If conflict increases at work, this means the team is not working in harmony, and something must have gone dead wrong, because people who are satisfied with their job and enjoy good communication do not fight each other.

Increased conflict at work is a sign of discontent; it can produce change for the better but it usually escalates and contributes to a disaster.

No conflict at work

No conflict at work is a very dangerous sign; it does not mean that all is well. On the contrary, it means all is bad but nobody wants to talk about it or take action to rectify it, for fear of something or someone. After a short period, the situation usually blows up and a disaster situation develops.

Lack of information

An important characteristic of a good leader or manager is the spread of necessary information to team members, so they can perform their tasks properly. Lack of firsthand information results in team members taking decisions relying on their own limited information, sometimes causing a disaster. That is why team meetings must be properly planned and successfully conducted to provide everyone with information and motivation, to safeguard company interests.

Decision-making process

Involving team members in the decision-making process insures they are committed to the team vision, mission, goals, values and expectations.

Failing to do so may have grave consequences, since everyone will be working on their own and not in harmony with the others.

Conflicting responsibilities

A ship with two commanders will definitely sink, and a job with two overall managers will usually fail. The principle of fixed responsibility states that, for a given period, an individual will accomplish the task when the responsibility for the completion of the task is fixed upon that individual.

Inadequate communications

Some of the main characteristics of a successful team are that they are a highly communicative group of people, they can discuss everything freely, and they have a crystal clear idea about team mission, vision, goals, values and expectations.

Team meetings must be planned and conducted to provide everyone with a chance to present their views and ideas about the work they are performing, and to clarify any queries they may have. The availability of proper communication channels insures that no real problem develops into a disaster; this is also in the best interests of the telecommunications company.

Rejecting performance evaluation

Rejecting performance evaluation systems has a detrimental effect on the operation of the organization. How can we insure operations are being carried out in the proper manner when we know nothing about the performance figures? How can we insure no disaster will happen when we have no information about the transmission quality of the common carrier, for example? If we do not measure the operating parameters of a broadcasting station, how can we know that the power output tubes need to be changed, otherwise they will fail in service, creating a disaster. Performance evaluation can be carried out on managers, engineers and technicians, as well as on equipment.

Resistance to change

People hate to change; psychologists argue that the only way to change someone else is by subjecting them to a psychological shock. Nevertheless, we can produce some change in others by setting an example ourselves. A more effective method is to select the right person for the job. When we recruit a faultfinding technician, they must possess some basic skills such as analytical deduction and critical thinking. If the person we already have does not possess these skills, it will be a waste of time and money to offer them training. Instead of trying to change people, it is better to assign them to positions for which they have the basic talents and skills.

Then training will be useful in raising their standard of work; this will be a beneficial and cost-effective decision.

4.1.3 Stages in the development of a disaster

The core

The core of the disaster is the main reason for the disaster.

The environment

The environment can stimulate and encourage the development of the disaster, due to unhealthy conditions, such as:

- A prevailing sense of carelessness

- Broken relations between management and workers

- Management bureaucracy

- Widespread corruption within the company

- Nobody cares for the welfare of the company

The catalysts

When the decision maker is not in direct contact with the different elements of the company, they have a tendency to become isolated, taking their information from nearby catalysts (a secretary or close associates). Here the decision maker risks failing to detect the warning signs of a disaster.

Ignoring warnings

Ignoring the warning signs means the crisis will develop into a full-blown disaster, unable to be stopped without losses to the company.

Tension and anxiety

Leaving the crisis without confrontation will have adverse effects on everybody in the organization, since everybody will be expecting some action to be taken on the part of the management. If no decision is taken, we may arrive at the point where people change their attitude from anticipating a disaster to watching it happen.

The breaking factor

The breaking factor is a minor action that triggers the disaster. The action itself might not be that important, but its effect is great because it happens when a crisis is waiting for something to turn it into a full-blown disaster.

The disaster erupts

The disaster is in full swing; here are some forms it may take:

- Interruption of service

- Blocking of some routes

- Fire in the building

- Hardware or software problems

- Delay in call connection

- Increased maintenance time

- Frequent and lengthy failures

- Decreased transmission quality

- Reduced company profits

- Increased employee turnout

- Increased customer complaints

- Loss of market share

4.1.4 Ways to categorize disasters

Chronological stage

The following stages should be studied carefully to come out with a proper plan covering what to do in case of a disaster. In general, this and other experiences should be compiled as case studies in the disaster management manual of the organization.

The early (infancy) stage

Here the newly born disaster begins to appear; one feels there is something strange going to happen but without knowing exactly what, when, how or where. One might notice, for instance, that the number of quick fades on a microwave link is increasing; this is normal and is taken care of by the frequency and space diversity equipment. Yet if this phenomenon has never been repeated before, the shift engineer should feel worried or at least begin to be alert, because something might happen, and in this case they should be prepared to take corrective action.

The growth stage

When the shift engineer in the previous example fails to detect the early warning signs (the increasing number of quick fades on a microwave link), the disaster grows bigger and bigger. Perhaps not only one link is experiencing the problem but several links working on different frequencies. Here the shift engineer couldn't fail to notice the developing disaster, although they can always ignore the warning signs.

Thus the shift engineer should be able to take some decisions to detect the real reason for this problem and try to avert an imminent disaster. This can be done by:

- Isolating the real reason for the problem; suppose it is the frequency, then what range is affected and can we change the antenna height to reduce the effect?

- Freezing the development of the disaster by eliminating its growth elements; maybe change the frequency of the spare link to the higher (or lower) frequency, or maybe examine the effect of changing the antenna height.

The maturity stage

It is rare that disasters will be left to develop until the maturity stage. But sometimes the decision maker (the shift engineer) is either very ignorant or very stubborn, so they will overlook or deliberately ignore all the early warning signs until the disaster develops to maturity and it becomes very hard to control. In the previous example all the links may fail to operate properly, and the bit error rate (BER) increases to the point where the quality of the transmitted data is out of spec. The link becomes practically inoperative, with huge losses to the organization and a tidal wave of customer complaints.

The decline stage

After disrupting the normal operation of the organization, the disaster weakens and its effect begins to decline. Yet this doesn't mean the disaster cannot develop again after some time. In the microwave example the fade pattern can change due to a change in climatic conditions, perhaps through a change in ambient temperature, humidity, pressure or wind velocity. Since all these elements are dynamic, the conditions conducive to fades may recur — there is the potential for another disaster.

The disappearance stage

Disappearance is the final stage in the development of a disaster. When all the elements responsible for a disaster lose their effect, the situation returns to normal. This does not mean the disaster is entirely a thing of the past. If it happened once, it can happen again.

Mode of occurrence

Mode of occurrence is one of the most important ways to differentiate the types of disaster, and it is also used in diagnosing them. It is astonishing that disasters happen in a repetitive way, and we have somehow learned to cope with them and minimize their effects. But strangely enough we usually make very little effort to stop them from occurring again. One way to counteract this passive behavior is by analyzing repetitive disasters and abrupt disasters.

Repetitive or cyclic disasters

If disasters occur repetitively, it usually means they have been handled in a way that eliminates the symptoms but which does not fully analyze the real causes. Repetitive disasters can be divided into problems happening at each stage of the country's economic cycle:

- Disasters that happen when the economy is in flourish:
  — Inflation problems
  — Shortage of raw materials
  — Shortage of capital
  — Shortage of technical staff
  — Shortage of administrative staff

- Disasters that happen when the economy is shrinking:
  — Problems in distributing the company's products
  — Problems of reduced profitability since all competitors are reducing their prices to get a larger share of the market in these difficult times
  — Decline in the actual pay the workers receive
  — Accumulation of finished products in the stores
  — Reduced utilization of the services of the telecommunications organization due to falling business activity across the country
  — Drop in the share price of the organization
  — Quick decline in the business performance of the organization

- Disasters that happen when the economy has hit bottom:
  — Unemployment problems
  — Organizational tension
  — Spread of arson

- Disasters that happen when the economy is reviving:
  — Operational problems: the organization has not yet reached its optimum performance
  — Staffing problems: the newly recruited personnel need to be trained and this costs money and time
  — Distribution problems: the distributors haven't the necessary expertise to handle the organization's products or services
  — Competition problems: many other organizations will enter the market once the economy begins to revive

Abrupt or random disasters

Randomly occurring disasters are very hard to predict, although some early warning signs are now available, e.g. for hurricanes and tornadoes, but no one has yet been able to give precise predictions of earthquakes. Rain, floods and lightning also have detrimental effects on the telecommunications industry, but very few early warning signs are available.

Here we can develop our plan on the assumption that if such a disaster happens then the agreed steps will be implemented accurately. In this case training is a vital factor in the success of the plan. If each and every employee knows their role exactly and has been well trained on the performance problems, we can be sure this disaster will be handled properly and the losses will be minimized.

Depth of effects

Disasters may have a superficial penetration into an organization or they may be much more deep-rooted.

Superficial disasters

Superficial disasters are disasters that do not constitute major interruption of the service; they do not cost the organization large sums of money. They are solved quickly by identifying the real cause and eliminating its prime mover. A cellular telephone company introduced a system where the customer bought an intelligent card (with a hidden identity number) and entered the hidden card number in the phone to use it for a specified number of calls within a particular period of time. The cards came in four denominations:

- A £50 card valid for 20 days for transmission and reception then 30 days more for reception only.

- A £100 card valid for 30 days for transmission and reception then 60 days more for reception only.

- A £200 card valid for 90 days for transmission and reception then 60 days more for reception only.

- A £300 card valid for 120 days for transmission and reception then 30 days more for reception only.

The vast majority of the company's customers found the £100 card the best, since they wanted to receive calls from their customers and at times make a phone call from places where there was no ordinary telephone. So the £100 cards were consumed very quickly and the customers had a very hard time buying new cards for their cellular phones. Immediately afterwards the £100 cards were found only in the black market at a higher price, sometimes 20% more.

Rumors flourished that the net revenue of the company was declining due to the large utilization of the £100 card and that this was the reason for 'withdrawing' it from the market. Complaints were filed with the responsible ministry, some lawyers even thought of taking the cellular company to court. More rumors went on that the intelligent cards in the black market had a validity date. This was not true, but it only served to raise the popular tension against the cellular company.

The company finally ordered the bulk manufacture of £100 cards and things went back to normal. This happened after a few months, during which the company's credibility declined in the eyes of its customers. The company had to launch a huge marketing campaign that cost it a lot. All these losses could have been saved if the early warning signs (high rate of card utilization) had been detected from the outset and taken care of immediately.

Deep-rooted disasters

Deep-rooted disasters are the most dangerous that can happen to an organization. A good example is a virus attack that disrupts the switching equipment software, resulting in network failure and loss of traffic revenue.

Strength of the disaster

Catastrophic, powerful

A catastrophic disaster will affect the whole existence of the organization, e.g. an earthquake or flood. Labor disputes and industrial actions can sometimes be catastrophic

Moderate

Moderate disasters can be handled if the disaster management team is ready and well trained. They may be hard to cope with from the organization's viewpoint, but their effects on the customers may be mild and can easily be handled, e.g. a virus that disrupts the billing file. This situation can be considered by the organization as catastrophic, but its effect on the customers can be minimized by sending the customers a bill containing the average of the last six months. When the system has been repaired, send them next month's bill containing the appropriate adjustments.

Mild, easily confronted

Mild disasters may be easy to cope with from the organization's viewpoint, but they may go unnoticed by customers. If we return to the billing virus, suppose it hit the billing records at the beginning of the month. In this case the situation could be rectified during the next week of the same month, then the customer would not notice any problems.

Scope of the disaster

General (macro)

A general disaster hits the core of the community or the country itself, e.g. arson activity within the telecommunications organization as part of a general campaign to overthrow the government.

Sectionalized (micro)

A sectionalized disaster does not spread outside the company. This kind of disaster has many types and takes many forms. Yet if its effects are not controlled or improperly taken care of, sometimes they may spread outside the organization.

Subject of the disaster

Technical

The system fails to provide the service as anticipated. A television transmitting station fails when the president of the republic is presenting the state of the nation address. A telephone exchange fails during the busy hours of the day when all business transactions are being carried out. The power supply from the generating station is interrupted while a surgeon is performing critical surgery, and so on.

The failure of these systems will be due to the failure of a subsystem or some components in the system. It is the duty of the O&M staff to predict and anticipate the time when a component is due for change, before it fails in service and generates a disastrous situation with its associated loss to the company.

Financial

All companies present themselves to the outside world through the annual report and accounts, which they distribute to their shareholders. These documents are supposed to reveal the true financial situation of the company. A company may fail due to the financial problems it encounters. This cannot happen abruptly or suddenly, as in the case of technical breakdowns; it takes more time, with many triggering events and warning signs. If these warning signs go undetected or ignored, or if the accounts are manipulated to cover them up, then one day the company will collapse. There are several key factors (or financial ratios) to consider:

- $\text{Activity} = \dfrac{\text{net sales}}{\text{invested capital}}$

  where invested capital = equity + long-term loans

- $\text{Profitability} = \dfrac{\text{net profit}}{\text{tangible net worth}}$

- Liquidity $= \dfrac{\text{current assets} - \text{inventory}}{\text{current liabilities}}$

- Capital structure $= \dfrac{\text{tangible net worth} + \text{long term loans}}{\text{fixed assets} + \text{net working capital}}$

It is a legal requirement that the company accounts should be audited. The auditor is an outsider who takes a very close look at the company, reveals any discrepancies, and gives warning signs in their comments on the account. The bank manager whose bank has a vested interest in the company is another party interested to insure the financial accounts reveal the true situation of the company. See Chapter 8 for more details.

Administrative

Administrative problems can cause disaster; some of the major causes of concern were given in Section 4.1.1 and on pages 113–117.

Legal

Legal problems can cause disasters too; conflict at work may rise and legal action may be taken against the culprits, then they resort to arson and a disaster occurs. Failure of a company to pay its customers the necessary compensation for their losses due to service interruptions can develop into a form of disaster; it can affect the financial situation of the company and also its credibility.

Combined

A combination of the above factors will only aggravate the problem, increasing the imminence of disaster.

Level of impact

Global disaster

Global disasters affect the whole world, e.g. the interruption of international telecommunications traffic carried by satellite communications due to a fault in one or more of the satellites carrying the service. Although there are alternative routes that can be activated immediately, they rarely carry the full load and it is always a problem to handle excess (overflow) traffic.

Country disaster

Country disasters affect a whole country, e.g. the interruption of a national gateway exchange that carries international telephone traffic to the world. Although alternative routes can be activated immediately, they rarely carry the full load and it is always a problem to handle excess (overflow) traffic.

Organization disaster

Organization disasters affect a whole organization, e.g. industrial action organized by a union demanding a hefty salary increase.

Department disaster

Department disasters disrupt a single department but their effects could spread to produce an organization disaster.

Section disaster

Problems within sections are rarely classified as disasters unless they show signs of spreading to other sections or departments.

4.2 ANALYZING THE ORGANIZATION

Planning and establishing a solid disaster management program is proactive in nature; it is undertaken to establish and implement capabilities for coping with disasters long before they occur. Here is an outline of the process.

4.2.1 Assessing disaster management capabilities

The first step is to check what we have already, our resources and inventory. We have to investigate if we can handle the probable disasters that might occur in the organization? If so, to what extent? Table 4.3 may prove useful in this respect:

- Effects (and losses) can be graded on a scale of 5, where 5 means major effects (and enormous losses) and 1 means minimum effects (and negligible losses).
- Coping facilities can be graded on a scale of 5, where 5 means excellent coping facilities are available and 1 means very poor facilities are available.
- Company preparedness can be graded on a scale of 5, where 5 means excellent preparedness and 1 means very poor preparedness.

4.2.2 Assessing preparedness to cope

Table 4.4 illustrates the different elements that must be considered when assessing the organization's preparedness for coping with a disaster. Availability can be graded on a scale of 5, where 5 means excellent measures available, and 1 means very poor measures available.

Table 4.3 Assessing an organization's disaster management capabilities

	System involved	Major disasters envisaged			Coping facilities	How well prepared
		Type or priority	Cause	Effects or losses		
1						
2						
3						
4						
5						
6						
7						

4.2.3 Assessing strengths and vulnerabilities

Table 4.5 is very helpful when assessing the organization's disaster management strengths and vulnerabilities. See Section 4.4.1 on SWOT analysis.

4.2.4 Stages of disaster preparedness

1. Identify the need for a disaster management program and obtain the management's consent.

2. Collect data about the available resources at the company that can be used in disaster prevention.

Table 4.4 Assessing an organization's disaster preparedness

Major anticipated disasters	Loss (US$)	Disaster avoidance systems	Availability	Needed funds (US$)
Interruption of local telephone service		Technical capability and staff expertise		
Interruption of international telephone service				
Other causes				
Fire in the generator room		Fire prevention, fire control and firefighting		
Fire in the exchange hall				
Other causes				
Flood in the basement		Flood detection and management		
Chemical spill from the battery room		Environmental contamination handling		
Hackers break into the system		Software security		
Mains power supply interrupted		Power backup facility		
Other major disasters		Related systems		

Table 4.5 Assessing disaster management strengths and vulnerabilities

Strengths	Vulnerabilities
1	1
2	2
3	3
4	4
5	5

3. Conduct a risk analysis exercise and rate the anticipated risks according to their effect on the organization's operations.

4. Identify disaster avoidance techniques needed to avoid and control the anticipated risks and reduce their effects.

5. Develop a disaster recovery strategy (or strategies) to deal with the anticipated risks.

6. Develop a system and network recovery plan.

7. Develop a system and network backup plan.

8. Develop a user recovery plan.

9. Develop a disaster management and service recovery training program.

10. Select and train the disaster management and service recovery teams according to the training program developed for each anticipated risk.

11. Test the recovery plan and conduct service recovery drills.

4.2.5   A model disaster management program

1. Conduct a risk assessment study.

2. Obtain management backing.

3. Select the program team and its coordinator.

4. Define the action plan; set tasks and deadlines.

5. Define expected outputs and presentation style.

6. Determine the budget to complete the work.

For the risk assessment study, conduct a preliminary analysis to obtain the risk of disaster and the corresponding cost to the company. Then compare your situation with some similar companies and gather information about disasters they have already faced and the costs they have incurred. These costs might be the direct cost of replacing equipment, missed sales opportunities, repairing damaged facilities, and revenue losses through reduced customer confidence.

Management sponsorship is a key element in the success of a disaster management program. Without it there is no hope of any meaningful progress. To obtain management sponsorship, you have to present a briefing to top management stressing the following points:

- The dangers the company is facing now, the dangers it will be facing in the near future, and the anticipated cost of loss.

- The dangers the industry is facing now and the dangers it will be facing in the near future.

- How other companies have dealt with the problem.

- The legal and regulatory aspects involved in the anticipated dangers.

- The estimated cost of the disaster recovery planning program and the anticipated gain to the company.

4.3   CONTINGENCY PLANNING FOR DIFFERENT SYSTEMS

The essence of a disaster management and contingency planing program is to identify the risks facing the organization and to propose measures that will insure the service will not be disrupted during a disaster; to prepare an emergency system and network to insure the continuation of all critical services during a disaster. Several tasks are required of the disaster management team:

1. Prepare a simple contingency plan since this will make its preparation simpler and its implementation easier.

2. Identify contingent events; concentrate on the negative events since they will have a detrimental effect on the organization.

3. Consider the necessary actions to cope with the negative contingent events.

4. Consider the available resources that can help in implementing the contingency plan.

5. Estimate realistically the extra resources and funds needed for implementing the contingency plan, and obtain management approval.

6. Put the plan together and recheck the implementation strategy to reveal any unexpected vulnerabilities which might necessitate some change.

7. Put the plan in writing and make sure it is easy to interpret.

8. Train you staff on how to detect the occurrence of the triggering event and on how to implement the plan properly.

4.3.1 Telephone and data networks

1. We will follow here the points detailed earlier and begin by preparing a simple, one-event contingency plan.

2. Let us consider that the triggering event in this case, revealed from the worst-case scenario, is fire erupting in the telephone emergency power supply room near the telephone exchange hall.

3. We next consider how fires can erupt in this place, the actions needed to insure they do not erupt, and the actions needed to fight any fire that does erupt. See Section 9.4.

4. Survey the available firefighting equipment and prepare an inventory.

5. Estimate the extra financial resources needed to buy extra firefighting equipment if the available items are not sufficient. Also check your firefighting personnel; have you enough staff or do you need more trained personnel? Obtain management approval for the plan.

6. Formulate the plan and check for any vulnerabilities. Use the vulnerability search, analysis and rectification method discussed in Chapter 6.

7. Put the plan in writing; explain the tasks clearly and unambiguously, to insure they cannot be misinterpreted. If more funds are needed then management consent is essential, but you may have to be very persuasive.

8. Training the staff to cope with the chosen triggering event (fire) is very important because the losses will be great if they do not act immediately, within the safe period before fire propagation.

4.3.2 Computer networks

1. We will follow here the points detailed earlier and begin by preparing a simple, one-event contingency plan.

2. Let us consider that the triggering event in this case, revealed from the worst-case scenario, is flooding in the computer room located in the basement of the company building. Flooding is rated as one of the major disaster-triggering events in the computer industry.

3. We next consider how a flood can occur in this place. Flooding can be due to natural causes, as detailed in Section 3.5.4. Also, flooding from sewage water can happen when drains are clogged in the premises (especially if this happened at night or during weekends and holidays). Consider the actions needed to insure flooding does not disrupt company operations, and the actions needed to safeguard the company against flood from natural causes or from internal sewage problems.

4. Conduct a survey of available information from the weather bureau and insure the updates are arriving regularly; this is very important. A survey of the sewage system is mandatory.

5. Estimate the extra financial resources needed to buy extra equipment or tools to cope with this disaster. Check your disaster management team; have you enough staff or do you need more trained personnel? Obtain management approval for the plan.

6. Formulate the plan and check for any vulnerabilities. Use the 'what if' analysis discussed earlier. You may need to transfer your computers to the first floor; this might seem expensive but it is probably less than the cost of flood-damaged computers.

7. Put the plan in writing; explain tasks clearly and unambiguously, to insure they cannot be misinterpreted.

8. Training the staff to cope with the chosen triggering event (flood) is very important because the losses will be great if they do not act immediately, before the flood actually happens.

4.3.3 TV and broadcasting networks

1. We will follow the points detailed earlier and begin by preparing a simple, one-event contingency plan.

2. Let us consider that the triggering event in this case, revealed from the worst-case scenario, is that the microwave link supplying the program to a transmitting station goes down.

3. We next consider how a microwave link can go down and stop functioning. In fact, several options are available:
   - The microwave link is composed of several hops, and one of them has gone down.
   - The power supply to a single hop link was interrupted.
   - A tornado erupted and damaged one of the microwave antenna towers.
   - The output stage in the microwave transmitter failed to operate due to a failure of the output stage transistor.
   - The air-conditioning equipment in the unattended microwave repeater station stopped operation and the temperature rise caused the transmitter to fail due to excessive heat.
   - The fuel tank supplying the main diesel generator in the unattended microwave repeater station became empty and the electrical power supply to the repeater went off. (There is no solar backup.)
   - There are many triggering events that can be listed; the worst-case analysis should determine the most relevant cause and the most suitable plan.

4. Survey the available resources and prepare an inventory.

5. Estimate the money needed to buy an extra fuel tank (if this is the chosen event). Or check the operating parametrs of the output stage in case a power transistor failure is the anticipated triggering event. Obtain management approval for the plan.

6. Formulate the plan and check for any vulnerabilities. You might find it is imperative to include more than one triggering event in this case. Use the 'what if' analysis discussed earlier.

7. Put the plan in writing; explain the tasks clearly and unambiguously, to insure they cannot be misinterpreted.

8. Training the staff to cope with the chosen triggering event is very important because the losses will be great if they do not act immediately.

4.3.4 Electrical power distribution

1. We will also follow the points detailed earlier and begin by preparing a simple, one-event contingency plan.

2. Let us consider that the triggering event in this case, revealed from the worst-case scenario, is lightning striking the electrical power substation supplying power to the telephone exchanges in the city.

3. We next consider how lightning can strike in this place, how the grounding system is made, the actions needed to insure lightning does not strike the power lines; we also consider actions needed to remedy the affected lines, or change the broken insulators, if a lightning strike has occurred. See Section 9.7.

4. Survey the available lightning survival requirements and prepare an inventory.

5. Estimate the money needed to buy extra earthing equipment if the available items are not sufficient. Also check your team personnel; do you have enough staff or do you need more trained personnel? Obtain management approval for the plan.

6. Formulate the plan and check for any vulnerabilities. Use the 'what if' analysis discussed earlier.

7. Put the plan in writing; explain the tasks clearly and unambiguously to insure they cannot be misinterpreted. If more funds are needed then management consent is essential, but you may have to be very persuasive.

8. Training the staff to cope with the chosen triggering event (lightning) is very important because the losses will be great if they do not act knowledgeably and immediately before, during and after the strike.

4.4 PROPOSALS FOR INTEGRATED APPROACH PLANS

4.4.1 SWOT analysis

SWOT analysis is a powerful tool used in assessing the organization; it begins with a survey conducted by the managers of the organization, taken from the viewpoint of each manager, to determine the following four elements:

- The strengths of the organization (S)
- The weaknesses of the organization (W)
- The opportunities facing the organization (O)
- The threats facing the organization (T)

The areas to be checked in each SWOT element will depend on the activity and discipline you are dealing with, but they can be broadly divided like this:

- Production
  — Facilities and capacities
  — Quality control costs and ratios

— Inventory capacity
— Production cost analysis and ratios
— Engineering costs and analysis
— Research and development expenses

• Marketing
— Total sales, and share of the market
— Customer views and degree of dependence
— Market growth rate
— Degree of competitive pricing
— Competitors

• Personnel
— Professional and technical staff
— Operating staff
— Management staff
— Sales staff
— Administrative staff

• Finance
— Cash flow
— Return on investment
— Profitability
— Invested capital
— Liquidity
— Capital structure
— Bad debts loss
— Accounting practices

They can be included in a form for completion by the managers. A proposed form is given in Table 4.6.

How to conduct a SWOT analysis

1. Analyze the response of the managers and compile a list of all absolute strengths, weaknesses, opportunities and threats facing the organization.

2. List the major strengths and weaknesses in each functional area.

3. List the major opportunities and threats facing the organization.

4. Discuss ways in which the company can profit from the opportunities open to it against the competition it faces.

5. Discuss ways in which the company can strengthen its position and remedy its weaknesses against the threats it faces.

6. Prepare a visualization of the four SWOT strategies.

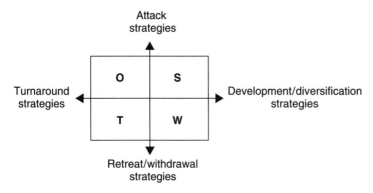

Figure 4.2   SWOT strategies

SWOT strategies (Figure 4.2)

- *Opportunities and strengths* point out the attack strategies to arrive at our objective.

- *Weaknesses and threats* require retreat and withdrawal strategies to prepare for a future attack to arrive at our objective.

- *Opportunities and weaknesses* indicate that we have to use turnaround strategies to arrive at our objective.

- *Threats and strengths* indicate that we have to adopt a development and diversification strategy to arrive at our objective.

4.4.2   Vulnerability rectification

Vulnerability rectification of weaknesses and threats identified in the SWOT analysis is very important because:

- It reduces the probability that a disaster will occur.

- It reduces the level of impact of this disaster if and when it actually happens.

- It reduces the consequences for the organization.

Table 4.7 is useful in identifying the rectification status and the action needed for each type of vulnerability.

4.4.3   The integrated approach plan

In many cases an organization will prepare several disaster management and service recovery plans, each of them covering individually and independently one or more of the organization's activities, without giving consideration to the need for a comprehensive disaster management and service recovery plan for the whole organization.

Putting into action several disaster management plans not only constrains the flexibility of the organization to adapt to change, but also reduces the overall functional reliability of the organization, since these complex plans might interfere with each other and pose a threat

Table 4.6 SWOT questionnaire

SWOT Questionnaire

Please indicate the competitiveness of the organization in the following areas:

- Production
  - Facilities and capacities
  - Quality control costs and ratios
  - Inventory capacity
  - Production cost analysis and ratios
  - Engineering costs and analysis
  - Research and development expenses

- Marketing
  - Total sales and share of the market
  - Customer views and degree of dependence
  - Market growth rate
  - Degree of competitive pricing
  - Competitors

- Personnel

- Professional and technical staff
  - Operating staff
  - Management staff
  - Sales staff
  - Administrative staff

- Finance
  - Cash flow
  - Return on investment
  - Profitability
  - Invested capital
  - Liquidity
  - Capital structure
  - Bad debts loss
  - Accounting practices

Please indicate which of these parameters can be included in each of the four elements of the accompanying SWOT table.

(*continued overleaf*)

Table 4.6 (*Continued*)

Strengths	Weaknesses
1	1
2	2
3	3
4	4
5	5

Opportunities	Threats
1	1
2	2
3	3
4	4
5	5

Table 4.7   Vulnerability rectification

Vulnerability			Rectification status[a]	Needed action[b]
Type	Where it exists	Its effects		
		Death		
		Injuries		
		Interruption of service		
		Loss of service		
		Loss of data and loss of Information		
		Damage to buildings		
		Loss of customers		
		Other effects		

[a]Rectification status can be graded on a five-point scale, where 5 means excellent system available and 1 means very poor preparedness requiring immediate rectification.
[b]The actions needed to rectify the vulnerability; an estimate of the cost is recommended.

to the overall functional continuity and recovery during and after the disaster.

An integrated disaster management and service recovery plan must be put as one integrated plan, not as the integration of several separate plans.

SUMMARY

To prepare a disaster management and service recovery plan, we must first identify the disasters we are facing. We have to analyze why the disaster occurred or might occur, and catalog the warning signs: technical, financial, administrative, behavioral and organizational. We have to identify the stages in a disaster and the different types of disaster, considering their formation, occurrence, depth, strength, scope, subject and level of impact. Next we have to examine the organization to assess its disaster management capabilities, preparedness, strengths and vulnerabilities. The contingency plan can then be formulated for the specific activity of the organization.

Organizations often prepare several disaster management and service recovery plans instead of one comprehensive plan for the whole organization. This constrains flexibility and reduces functional reliability within the organization. Interference between separate plans could threaten continuity and hinder disaster recovery. Formulate one integrated plan from the outset; do not try to integrate several separate plans.

REVIEW QUESTIONS

1. Define the following terms: planned disasters, contingency plans, risk assessment, integrated approach plans.

2. Give some examples from your work experience of a disaster that you identified using the explanations given in this chapter.

3. Explain how you would think methodically about identifying the different types of disaster that can happen at your workplace.

4. For each type of disaster you identify at your workplace, give your ideas about how it could be avoided.

5. What are the benefits of conducting a risk assessment at your workplace? How might you persuade the management to believe your findings?

6. What is meant by a SWOT analysis? How might a SWOT analysis benefit your organization?

7. How would you estimate the total cost of initiating a disaster management facility at your workplace? Do you think the cost would be justified? Give an example to show how the total cost of a disaster was far more than the estimated cost of a system to prevent it.

5

Designing for Disaster

OBJECTIVES

- Answer questions at the end of the chapter.
- Discuss the design options for coping with disaster.
- Give examples of equipment reliability considerations at your workplace.
- Describe network configuration and management techniques at your workplace.
- Analyze problems with colleagues or subordinates at your workplace.
- Suggest new ideas for coping with disaster at your workplace.

5.1 CHOICE OF TECHNOLOGY

5.1.1 Digital versus analog

Advantages of digital switching

- Digital exchanges have lower initial costs and lower annual charges than analog equipment. Maintenance staff savings are significant and interruption losses are minimal.

- Digital exchanges in a fully digital environment take up far less space than analog exchanges; it is normal to assume a ratio of 1:10 reduction in the needed floor area.

- Digital exchanges are equipped with traffic measurement equipment and traffic-handling routines that considerably reduce traffic congestion. Also there are many customer-related facilities that reduce the amount of time and the level of staff to perform many tasks, such as changing a subscriber's number.

- Transmission improvements have accompanied the change from frequency division multiplexing (FDM) to time division multiplexing (TDM) then to wavelength division multiplexing (WDM), combined with the move from two-wire to four-wire devices, and finally the change to dense WDM (DWDM). This has significantly reduced losses without requiring huge investment in new cable plants for the local distribution network.

Pros and cons of digital transmission

Advantages

- Lower cost is achieved by using common circuitry, since the digital telephone transmission and switching equipment uses the same kind of logic circuits as in computers. Mass production has dramatically reduced component costs while their capabilities are steadily increasing. As a result, a digital system costs less than a comparable analog system.

- With digital switching and transmission it is possible to dispense with some circuits (e.g. two-wire to four-wire conversion), eliminating hybrid and other analog-related equipment.

- Digital signals are easier to multiplex than analog signals; the channel separation filters are simpler to manufacture.

- Digital channel supervision is simpler and cheaper. The on-hook/off-hook signal is a train of binary 0's and 1's that can be represented with another bit in the data stream.

- Since pulses of well-defined and uniform rectangular shape represent binary signals, they are easy to reconstruct even if they are badly distorted by noise.

- Crosstalk is eliminated. There is no chance of overhearing a conversation on another circuit. Digital signals are highly resistant to crosstalk; if it does occur, the result is random noise — completely unintelligible.

- Digital systems can easily handle other signals besides speech, e.g. data or TV signals.

- Packet switching has increased the circuit utilization by a factor of 40% or more.

Disadvantages

- A digital link is connected to an analog network through a digital-to-analog convertor (DAC) in one direction and an analog-to-digital convertor (ADC) in the other direction. This setup is required at each and every interface between the two systems.

- Standard integrated circuits cannot tolerate the high voltage transients that might occur in a digital circuit, and their performance degrades with high temperature. Without careful design, this can lead to high failure rates.

- The DC and ringing current in the present-day telephone system must be separated from the digital logic circuits used in the digital system. The circuit that provides this service is termed BROSCHT (battery feed, ringing, overvoltage protection, signaling, coding, hybrid and testing). When the system becomes fully digital, using digital telephone sets, this circuit will be eliminated.

5.1.2 How technology affects disaster management

Going digital has great advantages, yet the increased transmission capacity does mean that a failure is very expensive and there are three other disadvantages:

- The expertise of the faultfinding team must be very high, so the cost of hiring technicians will be high, although the number of repair technicians per 1000 telephone lines drops dramatically by digitalization.

- Since a failure will involve many circuits (almost one million in the case of satellites), the cost of increasing the system reliability will rise dramatically. This normally means that redundant systems are used extensively, increasing the capital cost of the system accordingly.

- Highly trained people will be required in developing countries, but they can seldom be found since their own citizens will need many years of training to attain the necessary expertise. Developing countries who buy state-of-the-art telecommunications equipment are forced to recruit expatriate staff at very high cost.

5.2 FACILITIES AND BUILDING DESIGN CONCEPTS

The aim of any disaster management plan is to safeguard persons and property. When looking at facilities and buildings we have to consider three related aspects: the company site, the occupied building and any business issues.

In discussing risks within these three aspects, we have to classify the action into one of four categories (Section 6.3):

- *Immediate*: if the risk is imminent and quick action must be taken to avert a disaster.

- *Investigate*: when there is a potential problem that should be investigated to decide on the necessary action.

- *Analyze*: when the symptoms or interrelations of the risk are not very clear and they need more analysis to detect and decide whether or not a potential risk is involved.

- *Defer*: when there is no clear evidence of any potential risk symptoms.

5.2.1 Site-related issues

There is very limited opportunity to change any conditions of the company site. The only possibility is to check the site well before moving in; if it is not appropriate then the company should look for a new site. If the site the company already occupies has some clearly identified hazards then some action should be taken. One option is to move away if the risks threaten the company with high losses. Another option is to try reducing the level of risk from some dangerous events by providing proper disaster recovery strategies and resources. Here are some of the points that should be considered:

- There is sufficient parking and maneuver space for the fire brigade.

- The roads leading to the company can withstand the weight of the fire trucks.

- All high- or medium-voltage lines passing over the site are at a reasonably large distance from the highest building on the site.

- All structures on the site are properly protected from lightning.

- There is an ample water supply for use by the fire brigade in nonelectrical fires.

- The site is out of the flood stream passage.

- The site area is not threatened by hurricanes.

5.2.2 Building-related issues

General considerations

- The locations of the fire brigade pipes should be clearly marked.

- Fire drills should be conducted at proper intervals.

- There should be a contingency plan for the building.

- The disaster management team should be trained in how to perform their tasks within the contingency plan.

- The building directory at the entrance should clearly indicate the emergency and alternative exits.

- All employees should wear a special badge that is hard to counterfeit, so any intruder can be identified immediately.

- All visitors should be required to wear a visitor badge.
- There should be a system to register (and check) all who work after hours in the building.

Building basement

- The alarm should be audible in all areas of the basement.
- Fire detectors should be fitted in risk areas.
- Drains should be clear, unclogged and fitted with backflow protection.
- If there are sump pumps, they should be checked regularly for proper operation.
- If records are stored in the basement, they should be safely stored on shelves. So if a flood happened, or even if the drains clogged and water covered the basement, they would not be affected.
- If some confidential work is conducted in the basement, the windows should be properly covered and protected.
- If there is a possibility of fire erupting due to smoking, enforce regulations to forbid smoking in the basement.

Building floors

- Proper firefighting equipment and extinguishers should be available on each floor.
- The wall and ceiling material needs to withstand fire for a specified time before collapsing.
- The air quality should be checked regularly so the risk of choking becomes remote. Air quality monitoring equipment must be used.
- The air circulation needs to be designed so that air will be changed regularly, especially in the battery room, the computer room and the laser printer room.
- The direction leading to the main escape route should be clearly marked.
- The direction leading to the alternative escape route should also be clearly marked.

Building roof

- Proper firefighting equipment and extinguishers should be available.
- The roof material needs to withstand a fire on the roof or on the floor directly beneath.
- The roof material should have no cracks or apparent wear.
- The water tanks on the roof should not leak, and the drains on the roof should be clear and unclogged.

- The alarm should be audible in spite of the high noise of the central air-conditioning cooling fans (if they are used).

- There should be proper lightning protection.

Aisles, corridors, stairs, escape routes

- Aisles, corridors, stairs and escape routes should be properly fitted with lights from the mains supply and from a backup battery in case the mains supply is interrupted during a disaster.

- The main escape route and alternative escape routes should be clearly marked, and they should be wide enough to accommodate the rush that occurs during a disaster evacuation.

- Aisles, corridors, stairs and escape routes should always be kept clear and unobstructed. Path site bands help in the quick evacuation of employees.

Building elevators

- Elevators should have automatic lobby recall when there is a fire alarm.

- There should be a plan for rescuing anyone trapped inside an elevator during a fire alarm.

- Every elevator should be fitted with brakes to stop it immediately when the ropes are accidentally cut in a disaster.

Special rooms

Telephone exchange, TV studio, computer room

- Air circulation and air-conditioning are appropriate.

- Smoke, heat, moisture and fire detection equipment should be fitted.

- Firefighting equipment should be available and staff should be trained how to use it.

- No storage of corrosive material should be allowed.

- All equipment should be properly secured so it can do no harm during a moderate earthquake or hurricane.

- No plants should be allowed in these rooms.

Battery room

- There should be no apparent signs of acid or corrosion in the battery, there should be sufficient ventilation and the room should be equipped with fire extinguishers.

- Air circulation is appropriate.

Electrical risks

- The company should have a mains power supply together with an emergency supply or an emergency generator that can handle the critical load.

- The emergency power supply must be located outside the building (in case of fire) in a place immune from flooding; it should be regularly tested for automatic starting and proper operation.

- The fuel tank for the emergency power supply should be securely located outside the building, accompanied by proper firefighting equipment.

- The incoming power should be monitored regularly for quality, spikes and harmonics.

- Wiring and cabling circuit diagrams should be up to date and they should indicate the present state of the circuit.

Fire risks

Section 9.4 considers basic concepts when implementing a fire protection plan. Here are the main objectives for reducing the risk of fire:

- To reduce fire incidence

- To control fire propagation and spread

- To protect personnel and equipment

- To provide means of escape for trapped persons

5.2.3 Business-related issues

Business-related issues concern the conduct of business processes to insure the proper environment is available for professional work. The relevant equipment, systems, procedures and operational practices will be covered in later chapters.

5.3 EQUIPMENT RELIABILITY CONSIDERATIONS

Reliability can be defined as the probability that a system or a piece of equipment in the system will perform its required function, under previously specified conditions, for the required period of time. Reliability is in fact quality during a specified period of time. Alternatively it can be defined as the probability of nonfailure during a specified period of time. Failures may sometimes bring disaster, so insuring high reliability is a good approach to disaster management.

5.3.1   Reliability and cost-effectiveness

The price–performance indicators of the telecommunications sector, now a trillion-dollar industry, have shifted it to the number three position in the global economy. This is coupled with larger and more complex communications systems. But due to the high cost of failure, in terms of repair and lost revenue, the telecommunications sector has focused on achieving and maintaining high reliability standards.

Consider a faulty communications satellite carrying say 500 000 international telephone circuits. If the revenue of each circuit is taken as US$ 1.0 per minute, revenue lost per minute amounts to US$ 0.5 million. To repair such a satellite means a mission into space, at very high cost. The reliability of the satellite must therefore be greater than 99.999%, probably the highest reliability of a manmade system.

A reliability–cost curve (Figure 5.1) illustrates the relationship between cost and reliability.

At the beginning of the curve, the low reliability (low cost of achievement) means the cost of repair will be very high; at the end of the curve, the cost of failure is low and the cost of achieving this high reliability is excessive. The best bet is to aim for the intersection of the two curves, which gives the lowest total cost and the optimum cost reliability. Thus, every piece of equipment should be evaluated in terms of how much it costs to achieve versus any savings that will be gained.

The well-known bathtub curve (Figure 5.2) illustrates the failure rate during the operating life of the equipment.

It has three distinct regions:

- *The early failures period* is sometimes called burn-in or infant mortality. Here the failure rate is high due to low standards of manufacture and quality assurance, but the rate decreases steadily with time.

- *The constant failures period* is sometimes called the useful life, stress-related failure or random failure. Failures may be caused by substandard components or by severe operating conditions, beyond the

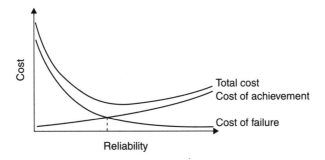

Figure 5.1   Reliability–cost curve

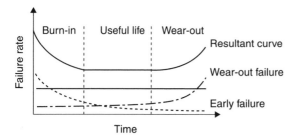

Figure 5.2   Bathtub curve

Table 5.1   Regions of the bathtub curve

Region	Failure characteristics	Reasons for failure symptoms
Burn-in	High failure rate that decreases steadily with time	Substandard manufacture and quality assurance
Useful life	Constant but low failure rate	Stress-related failures due to random fluctuations of stress transients exceeding the maximum design figures
Wear-out failures	Low failure rate that increases steadily	Time-dependent failures such as oxidation, corrosion, breakdown of insulation, normal wear or fatigue

design values. The useful period (constant failures period) can be somewhat extended by proper maintenance and overhaul programs.

- *The increasing failure rate period* is sometimes called wear-out. Failures occur through time-dependent factors such as corrosion, insulation breakdown, normal wear or fatigue.

The failure characteristics and their related reasons are given in Table 5.1.

5.3.2   Reliability in the system's life cycle

Here are some relevant definitions. Study them carefully so you understand them; we will need them later on.

Reliability function

The **reliability function** indicates the probability that a system will be successful for at least a specified time t. It is defined as

$$R(t) = 1 - F(t) = \int f(t)\, dt$$

where $F(t)$ is the unavailability function, so $1 - F(t)$ is the availability function, and $f(t)$ is the time to failure expressed as an exponential density function, since as the time period becomes longer the probability of failure becomes higher. The function $f(t)$ can be expressed as $f(t) = (1/\theta)e^{-t/\theta}$ where θ is the mean life and t is the specified period of time.

Failure rate

The failure rate is the rate at which failure occurs during a predetermined period of time. If λ indicates the instantaneous failure rate then the failure rate per hour can be expressed as

$$\lambda = \frac{\text{number of failures}}{\text{total operating hours}}$$

Mean time between failures

The mean time between failures (MTBF), symbol θ, is the mean value of the length of time between two consecutive failures, expressed as the ratio of the total cumulative time to the total number of failures. Thus $\theta = T/K$ where T is the cumulative time and K is the number of failures. Most of the time we assume constant failure rates. Then we can write $\theta = 1/\lambda$, which means the failure rate and the MTBF are connected by a simple expression.

Mean time to fail

The mean time to fail (MTTF) has the same meaning as MTBF, except it is used for systems or items that cannot be repaired, such as transistors or ICs. MTBF is applied only to systems or components that can be repaired.

Mean downtime

The mean downtime (MDT) indicates the average duration of a system interruption. It is usually expressed in percentile terms, and it is used when the outage affects the system's reliability or availability.

Mean time to repair

The mean time to repair (MTTR) is usually expressed in percentile terms, and it is used to estimate the repair time. When it is stated that 90-percentile repair time shall be 0.5 hour, it means that only 10% of the repair activities shall exceed 0.5 hour.

Availability

Availability A is the preferred figure of merit for assessment purposes. It can be expressed in several ways, but the most important in our discussion is operational availability, given as follows:

$$A = \frac{\text{MTBM}}{\text{MTBM} + \text{MDT}}$$

Where MTBM is the mean time before maintenance and MDT is the mean maintenance downtime.

Reliability component relationships

Any communications system is composed of several components, so in studying the system reliability relationships, one must calculate the reliabilities of the different components. These can be connected in series, in parallel or in series–parallel.

Series system

The overall reliability of a series system composed of say three subsystems (Figure 5.3) is the product of the reliabilities of the individual subsystems:

$$R_T = R_a R_b R_c$$

If $R_a = R_b = R_c = R$ then the overall reliability will be $R_T = R^3$. Suppose $R_a = R_b = R_c = 0.9$ then the overall reliability will be 0.7299, which is lower than the reliability of each individual stage.

Parallel system

The overall reliability of a parallel system composed of say three subsystems (Figure 5.4) is expressed as follows:

$$R_T = 1 - (1 - R_a)(1 - R_b)(1 - R_c)$$

If $R_a = R_b = R_c = R$ then the overall reliability will be $R_T = 1 - (1 - R)^3$. Suppose $R_a = R_b = R_c = 0.9$ then the overall reliability will be 0.999, which is very high. If the system is composed of only two subsystems, the overall reliability can be expressed like this:

$$R_T = R_a + R_b - R_a R_b$$

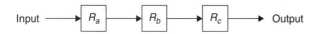

Figure 5.3   Series reliabilities

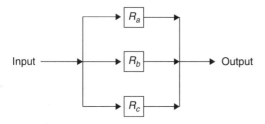

Figure 5.4   Parallel reliabilities

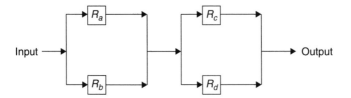

Figure 5.5   Series–parallel reliabilities

Series–parallel system

Consider the overall reliability of a series–parallel system composed of two series subsystems themselves consisting of two parallel subsystems (Figure 5.5). The overall reliability is given as follows:

$$R_T = [1 - (1 - R_a)(1 - R_b)][1 - (1 - R_c)(1 - R_d)]$$

If $R_a = R_b = R_c = R_d = R$, the overall reliability will be $R_T = 1 - (1 - R)^2)^2$. Suppose $R_a = R_b = R_c = R_d = 0.9$ then the overall reliability will be 0.980, which is still a high system reliability.

5.3.3   Achieving reliability

Design for reliability

Choice of parts

The quality standards of the chosen parts or components directly affect the resulting failure rate. Thus the proper selection of components from high-quality products will insure the system has high reliability.

Choice of technology

The type of component technology chosen for a communication system has a major effect on the system's reliability. The greater the sophistication of the technology, the greater the desire to provide many facilities. These added facilities lead to additional interfaces and higher component count, leading to higher failure rate and lower reliability. If we increase the reliability, the cost of the product will also increase.

Complexity

Adding more components or subsystems to a communications system reduces the reliability. If we have a system of three subsystems in series, each with a reliability of 0.9 (90%), then the system reliability will be $0.9 \times 0.9 \times 0.9 = 0.729$. If we add another subsystem of reliability 0.9, the system reliability will drop to 0.656. To keep the system reliability at 0.729, we have to increase the subsystem reliabilities to 0.923.

Derating

Derating means operating a component below its rated stress level (lower voltage, current, power, temperature, pressure, and other operational and environmental conditions) in order to obtain a longer life or a more reliable performance.

Environmental stress

Environmental stresses increase the failure probability of telecommunications systems. Here are some examples:

- High temperature
- Low temperature
- Thermal shock
- Vibrations
- Seashore atmosphere
- Dust
- Biological effects
- Reactive gasses
- Radioactive radiation

Redundancy

Redundancy can be applied in many ways; its effect is to increase the system availability because the failure rate has dropped (Section 5.3.2). Yet redundancy has many drawbacks, the most important being higher cost. Here are some of the other penalties:

- More space
- More weight is important for satellite payloads
- More component failures
- Common mode failures
- Lower ROI

Avoid construction failures

- Improper sealing of the equipment
- Poor design of the controls
  — accessibility
  — visibility
- Inadequate electrical design

- Poor performance of metals
  — seashore environment
  — humid conditions

- Inadequate insulation

- Poor performance at high temperature

- Poor cooling of sensitive components

- Poor wiring design

- Poor shock protection

- Poor vibration protection

- Poor mounting of components or parts

Burn-in, screening and maintenance

The burn-in process

During burn-in we operate the newly purchased equipment at elevated stress levels in order to accelerate any possible failures; this reduces the number of suspect components and increases the system reliability.

The screening process

We use screening to reveal any suspect component by performing visual, electrical and mechanical tests. These tests help to increase the system's reliability by revealing any weak item. Here are some of the tests:

- Visual inspection

- Temperature recycling

- Mechanical shock

- Vibration

- Thermal shock

- X-ray inspection

- Voltage and electrical cycling

Maintenance

Proper maintenance is essential; here are some of the options:

- *Routine maintenance*: follow the manufacturer's recommendations.

- *Preventive maintenance*: change components before they are supposed to fail.

- *Corrective maintenance*: change components after they fail; do not rely on corrective maintenance.

- *Reliability-centered maintenance*: plan your maintenance around component reliabilities.

- *Predictive maintenance or condition monitoring*: monitor symptoms and use the data to predict occurrence probabilities for failures and service interruptions.

A well-established maintenance program will help to:

- Eliminate unnecessary maintenance work

- Reduce maintenance rework costs

- Reduce lost revenue because of interruption

- Reduce the cost of spare parts inventory

- Increase transmission quality of service

- Extend the operating life of the telecommunications system

- Increase the traffic-handling capacity

- Reduce overall maintenance costs

- Increase overall profits

5.3.4   Reliability assessment and modeling

Reliability assessment is carried out to insure the system performance will satisfy the communications system's availability standards. This can be done in several ways, and each of them is possible for certain types of equipment. The success of one method compared with another depends on the amount of information available about the equipment under consideration.

Parts count

Parts count involves a simple addition of component failure rates without allowing for the different types of stress. This is a quick and easy process to perform, yet it provides only a worst-case scenario for the reliability.

Failure mode and effect analysis

Failure mode and effect analysis (FMEA) is a very powerful technique for enumerating the possible failure modes. This is achieved by tracing through the characteristics and consequences of each mode of failure on the system as a whole. Table 5.2 indicates one way to perform it.

Table 5.2 FMEA form

System		Subsystem		Unit		Subunit	
Circuit diagram			Prepared by			Date	
Item	Failure modes	Causes of failure	Effects of the failure	Failure rate (/million hour)	Criticality	Action needed	
Diode	Open circuit	Overload Environ-mental stresses Defective material Wrong type	Circuit interruption	0.025	Critical	Check operating parameters	
	Short circuit		Transistor shorted	0.015	Critical	Perform preventive, maintenance	
	High reverse current		Defective operation	0.060	Critical	Buy high-standard diodes	
Valves							
Resistors							
Capacitors							
Transistors							

In many disaster scenarios the failure of a component, as the initi-ating event, may have wide-ranging results; there might be a simple breakdown or there might be a major disaster. Although FMEA is a powerful tool, it becomes hard to implement when there are many components. Two variants of FMEA are event tree analysis and fault tree analysis.

Event tree analysis

Consider a command center for disaster control that should be working 24 hours a day, 365 days a year. It must not suffer any power interruption, so it has a mains power supply and a standby diesel generator, which operates when a no-voltage sign is detected by a voltage-monitoring unit. The worst-case scenario happens when the power is cut off. According to Figure 5.6a, the power will be cut in three cases out of four, due to a fault in the voltage monitor or in the diesel generator. This can be represented in the form of a fault tree diagram (Figure 5.6b). The fault tree diagram provides a graphic representation to help the maintenance engineer assess the impact, consequences and effects of the failure.

Fault tree analysis

Main supply cutoff can be due to

- Power supply interruption from outside the premises
- In-house power supply interruption, perhaps due to:

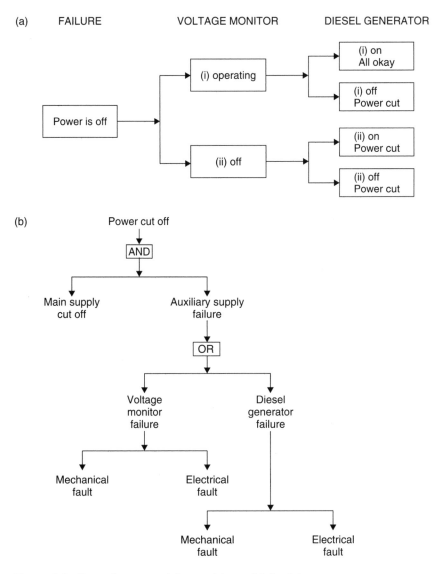

Figure 5.6 Tree diagrams: (a) event tree, (b) fault tree

— Mechanical fault in:
 • main switch parts
 • auxiliary switch parts
— Electrical fault in:
 • main switch parts
 • auxiliary switch parts
— Control circuit fault

Auxiliary supply failure can be due to

- Diesel generator failure
  — it fails to respond to the voltage monitor signal

- Voltage monitor failure
  — it fails to send a signal to the diesel generator

Simulation

Simulation caters for very complex systems; it uses computer capabilities to simulate all possible failure rates and modes.

The four methods compared

Method	Advantages	Disadvantages
Parts count	Simple and quick	Does not address the modes or stresses
FMEA	Addresses each component failure mode	Difficult to apply to complex systems
Fault tree	Flexible Addresses complex systems Can be computerized	Top-down process Some important configurations can be missed
Simulation	Addresses complex systems with any distribution of failure rates	Very expensive, especially for high-reliability systems

5.3.5 Software reliability and quality

In software failures there is no physical change which causes a working unit to cease functioning; in fact, software failures are errors that do not always become evident immediately, because of the complexity of the program. Also there is no wear-out feature, unlike the hardware bathtub features.

Software faults are usually called bugs; they arise as a result of some parts of the code being used for the first time, or due to corruption caused by some outside program, perhaps received over the Internet. The fault may not be discovered immediately, because some bugs stay dormant until a specific operation is performed, then they appear and disrupt it. A bug may lead to an error. An error may propagate and lead to a failure, if the system is not equipped with an error recovery program.

About 60% of software faults are committed during the design phases, the remaining 40% occur during coding. When designing a complex system, faults will most likely stem from ambiguities and omissions in the specification stage. Yet the major sources of faults are:

- Faults due to the specification:
  - — incorrect requirements
  - — inconsistent requirements
  - — incompatible requirements
  - — unclear requirements
  - — illogical requirements
  - — incomplete requirements

- Faults due to the design
  - — improper design approach
  - — unpopular design language
  - — improper design implementation

- Faults due to the coding
  - — semantic errors
  - — syntax errors
  - — logical errors

5.4 NETWORK CONFIGURATION AND MANAGEMENT

5.4.1 Modern network configurations

We have seen in Chapter 2 that any telecommunications system is required to transmit the communications information faithfully from its original point to its destination, and that the main components of a telecommunications network are:

- The terminal equipment, such as the telephone, telex, fax, computer, TV and radio.

- The switching equipment, responsible for establishing a connection between subscriber and destination.

- The transmission equipment for connecting subscribers to the network and exchanges to each other.

Network configuration techniques are used to set up a network that combines these three components and they are used to calculate the traffic flow in the network to establish economic viability. The network can be established in three basic formations.

The mesh network

Figure 5.7 shows a mesh network where all the exchanges are directly connected to each other. If the number of exchanges is n, the number of links C between exchanges is given by $C = \frac{1}{2}n(n - 1)$.

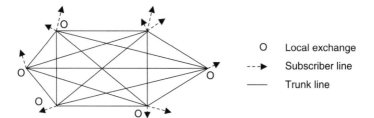

Figure 5.7 Mesh network

The star network

Figure 5.8 shows a star network where all the exchanges are connected to a transit exchange in a star formation. Here the number of links equals the number of exchanges, i.e. $C = n$.

The composite network

Figure 5.9 shows a composite network where all the exchanges are connected to a transit exchange, like in the star network; other exchanges are connected in a mesh formation.

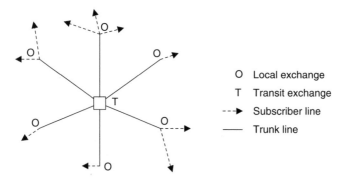

Figure 5.8 Star network

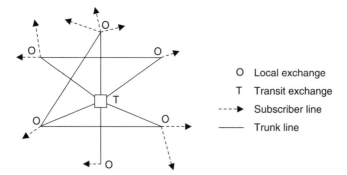

Figure 5.9 Composite network

Comparison between the three configurations

	Mesh network	Star network	Combined network
Transit exchange	Not required	Required	Required
Circuit efficiency	Low, inefficient traffic routes	High, traffic routes concentrated	Optimal
Effect of local exchange failure	Service will be provided through other parts if fully connected	Service will be disrupted to the failed destination	Service may be provided through other parts
Effect of transit exchange failure	Not applicable	Service totally disrupted to the network	Service may be provided through other parts
Cost estimates	Trunk network raises the cost	Transit exchange raises the cost	Optimal cost
Use	In areas with large interexchange traffic	In areas with low interexchange traffic	Optimal use

5.4.2 Problems in modern networks

Problems in modern telecommunications networks encompass many topics; one of the most important is how to keep old systems working in a modern network without degrading its quality of service. Old systems were intended to work properly for 10–15 years before replacement, but technological advances have made them obsolete before their working life is over. Thus the rate of return on investment for the telecommunications authority becomes very low. If the organization is able to interface the modern network with the old equipment, or even some of it, then it gains a return on its investment. Yet this strategy may greatly increase the probability of disaster.

Maintaining old equipment

Telecommunications equipment continues to serve in the network for a long time, long after new systems have been introduced. Old equipment will continue to function within the new system and contribute to the overall performance of the network.

Yet old equipment costs more, since it takes longer to maintain and fewer staff know how to service it. The mean time between failures (MTBF) is usually short. Thus, any efforts to reduce costs and improve reliability will be welcomed.

The interest in old systems is justified, since they are depreciated over 15–20 years, while new systems are introduced almost every day. If you throw away your old equipment, you lose a lot of money; if you keep it as it is, you also lose due to the anticipated higher rates of failure and the increased operation and maintenance costs. Here is the best solution:

1. Devise some automatic testing apparatus for the transmission and switching networks to cater for the old equipment.

2. Trace and solve persistent problems with the old plant.

3. Consider the possibility of selling the old equipment to another telecommunications company which doesn't have the resources to buy state-of-the-art technology.

Interfacing new and old

Older systems, even when they work economically, face a rather unusual problem of interfacing with new technology. The most obvious problems nowadays stem from the introduction of the Internet. High-speed transmission cannot be handled by the old copper outside plant, so telecommunications companies have been forced to begin a worldwide change to fiber-optic cables. The cables that are being changed could have served for 20 years more, but they became obsolete due to the introduction of new technology.

Keeping old plants economical

Although it may be working properly, this doesn't necessarily mean that an old plant is working economically, because the few failures it causes may cost the company a great deal of money. Conduct an optimum replacement analysis to decide when the old plant should be changed. Consider the following points:

- Total cost over the period (O&M costs plus consequential cost of failure)

- Resale value of the old plant

- Cost of the new plant

- Cost of the money (money borrowed from the bank)

5.4.3   The need for network management

Network management grew out of international operations, the most lucrative part of the telecommunications business. Two important management tools were developed to handle networks: common control in telephone exchanges and alternative routing of telephone traffic. The introduction of these tools increased operational efficiency, call-handling capacity and call completion rate. In developing the requirements of a network management system we have to consider the following key elements:

- Service aspects:
  - — rapid deployment of new services
  - — quicker service activation

- Technology aspects:
  - — efficient data management and data distribution
  - — elimination of physical overlay networks

- Business aspects:
  - — reduced operating costs
  - — competitive and timely services
  - — more flexibility in all aspects

5.4.4 Teletraffic engineering

Telecommunications networks are dimensioned to provide an agreed quality of service under normal load conditions. When the demand becomes greater than the network can handle efficiently, then congestion occurs, the call completion rate decreases and failure becomes imminent.

Teletraffic engineering helps to analyze the traffic profiles of each exchange in the network and predicts the pattern of change, then corrective measures can be applied to insure the peak demand is handled satisfactorily. Here are some of the measures:

- Temporary alternative routing (an expansive measure)

- Cancellation of alternative routing (a protective measure)

- Altering the alternative routing (an expansive/protective measure)

- Inhibition of routing to certain destinations (a protective measure)

- Directionalization of both-way-operated circuits (an expansive/protective measure)

- Circuit-busying techniques (a protective measure)

- Recorded messages that explain the situation (a protective measure)

- Increased efficiency of operator-assisted calls (an expansive measure)

5.5 THE PEOPLE

We always manage people not machines, yet few managers are trained to be *people managers*. Engineers need to be given good preparation to become professional managers. The relevant topics are motivation and conflict resolution. People are instrumental in disaster management; if the human relationships within the company are not healthy, a disaster becomes much more likely.

5.5.1 Motivation

Motivation is getting somebody to do something because they want to; manipulation is getting somebody to do something because you want them to. To manage people properly we have to begin by motivating them. Motivation is continually discussed but seldom understood. Remember that fear no longer motivates people in any free industrialized society. You get more of the behavior you reward, not what you hope for, ask for, wish for, or even beg for. So you must be the model for your people.

Prerequisites of motivation

- You have to be motivated yourself before you can motivate your team.

- Motivation requires a goal, so try to have a clearly defined objective.

- Motivation is achieved in two stages: find a goal then find out how to achieve it.

- Motivation cannot last forever; when motivation wanes, you need to encourage it again.

- People will work harder to get recognition; you can motivate people by praising them judiciously.

- Recognition may take many forms:
  — personal accreditation publicity
  — a change in job title
  — public praise
  — a status symbol
  — certificates and wall plaques

- Money is a very dangerous form of recognition; once motivated by money, people may not respond to other incentives and they may want large sums.

- Participation motivates people; try hard to make everyone feel part of the project.

- People are motivated by progress but only if they can see what they've achieved.

- Don't get into a fight if you know you will not win; losing rarely motivates people.

- Group identification motivates people by making them feel part of a team.

Demotivation: what not to do

- Don't undermine the confidence of your staff; reduce criticism and use constructive remarks or observations.

- Don't create a feeling of insecurity within the team. Indicate clearly what happens if a team member fails, but explain that it's not the end of the world.

- Avoid negative opinions; treat people with respect and motivate them sensitively.

- Don't create a feeling that your people are unimportant; this is a highly demotivating attitude.

- Avoid failures in communication. Never say, 'I don't know what's going on.' Try to put things more constructively: 'Let's see what the situation is now. How are we progressing?'

Incentive motivation

Contests and competitions are wonderful motivators, but you must set the rules properly. Here are the relevant points:

- Insure every participant has a chance to win, otherwise they won't try. Challenge only motivates if we can win.

- Long-term contests are of no value, three months is about the maximum duration.

- When setting a contest, ask yourself exactly what you want to achieve with it. You must have an objective.

- Prizes must be tangible, but they mustn't be money.

- Every contestant should know exactly what to do, what they could win and when the winner will be announced.

5.5.2 Conflict and confrontation

Conflict is a perceived divergence of interests, or a belief that the conflicting parties' current aspirations cannot be met or achieved simultaneously. Conflict can lead to disaster for the company or organization where the parties work; it may sometimes lead to arson. The reason for conflict is that we have the same interests, but they conflict. In this case we have to look out for, and endeavor to stress only our interests, not our position. Here are the objectives of learning how to deal with conflict and confrontation:

- To appreciate and accept human differences as inevitable.
- To help lose your fear of conflict.
- To learn techniques for dealing with conflict.
- To learn alternatives to the fight/flight reaction.
- To prefer solving problems over getting even.
- To be able to save your organization from disaster.

Dealing with disagreement

- Do an intellectual exercise with the other party
- Refuse to take the bait by keeping silent
- Agree to disagree
- Forgive the past

Pros and cons of conflict

Ironically enough, conflicts can sometimes serve the organization well. Here are some of their advantages:

- They can help to resolve discontent
- They can produce a change for the better
- They can lead to innovation
- They may change people's behavior

Conflict is bad when it escalates, and when it does not achieve any of these advantages.

Levels of conflict

Until conflict is resolved, it may escalate through five levels:

- *Accusations and threats*: only one issue is involved.
- *Proliferation*: more than one issue is involved.
- *Generalization*: conflict is generalized to the entire relationship.
- *Retaliation*: the parties involved move from self-interest to a desire to get even; the original issue may be forgotten.
- *Spreading*: other people become involved and the conflict spreads; it becomes very hard to solve and sometimes runs out of control.

Mr Zinga and Mr Bingo

Mr Zinga and Mr Bingo are business partners. Zinga has made a business transaction but Bingo is not happy with it. He tells his partner, 'I didn't like the way you sold the equipment for only $100 000. You could easily have got $150 000.' Bingo has begun the accusations.

Zinga might then have said, 'This was a tough bargain, next time you can do it yourself,' and the matter may well have been settled calmly. But Zinga has become irritated. He jumps one step forward and reminds Bingo of his own recent failures. Be thankful we have any funds for our cash-strapped company, that is Zinga's message. And he has now pushed the conflict to the second stage — proliferation.

Bingo is very upset. He jumps straight to level three: 'In fact, my dear Zinga, I never liked the way you handled all your transactions. You have made us lose enormous amounts of money.' This is a sweeping generalization and Zinga wants to get even. He has forgotten their original disagreement, he simply wants to retaliate. He tells his partner that their association has never been a success because Bingo himself is a total failure. The conflict has reached level four.

The furious Bingo immediately contacts all his business associates, giving them his side of the story and asking them not to do business with Zinga, his 'former partner'. He even attacks Zinga in the newspapers. The conflict has now spread to other people; it has escalated to the fifth and final stage. It is too late to back down; Zinga and Bingo are out of control.

Without pausing to discuss their thoughts and feelings, Zinga and Bingo rapidly went from level one to level five. And amazing as it may seem, this little story is remarkably true to life.

Conflict resolution

An interesting question is, How can we prevent conflicts and solve problems? It is hard to prevent conflicts since they are part of human nature, but we can reduce the rate at which they happen. We can use human relations, teamwork and motivation. Here are five strategies for conflict resolution.

Yielding

Yielding is a unilateral coping strategy that we use when:

- The other party's goals are more important than our own goals.
- Our relationships with the other party are unstable; we may be keen to keep a low profile in a new job.
- Approval from others is more important; we may want to give up the confrontation to avoid making our boss angry.
- We feel threatened for any other reason.

Yielding has three advantages:

- It is an excellent bargaining chip.
- It can save a great deal of time and hassle.
- It can prevent conflict escalation if the cost of escalation is more than the cost of giving up.

And three disadvantages:

- It produces very low joint benefits for the two conflicting parties.
- It rewards intimidating behavior by the aggressive party.
- Regret develops later on for the yielding party.

Withdrawal

Withdrawal means escaping from the conflict either physically or psychologically; it happens in two ways: breaking off voluntarily and giving up involuntarily. Withdrawal occurs in the following cases:

- If the persons involved have high concern for themselves and very low concern for the other party.
- If the persons involved have other alternatives.
- If the persons involved feel very angry and spiteful, so that they may give the other party the cold shoulder.
- When the persons involved are feeling threatened, because they think the conflict will escalate.

Withdrawal has three advantages:

- It saves time and nerves, and it diminishes the sense of frustration.
- It forces people to develop better alternatives.
- It can mean buying time or adopting a cooling-off strategy — a temporary withdrawal.

And three disadvantages:

- The conflict still exists, it does not get resolved.
- Withdrawal has very low joint benefits for the persons involved.
- It causes high levels of frustration for the withdrawing person.

Inaction

Inaction, a noncoping strategy, occurs when people have very low concern for their own goals and the other party's goals, when they are afraid of conflict, or when people don't want to rock the boat. The reasons for inaction are procrastination and denial:

- Procrastination means involving yourself in another activity of lower priority, so you become convinced you are busy and have no time to take action.

- Denial means convincing yourself there is no problem to begin with; denial is more dangerous than procrastination.

Inaction has three advantages:

- When you think that time will take care of the problem.

- When the issue of conflict is sensitive, e.g. if one of your colleagues has bad odor.

- When you think the conflict will escalate; this reason is always used as an excuse for not facing conflict.

And four disadvantages:

- Hostilities grow faster beneath the surface.

- The problem is compounded and conflict escalates.

- No joint benefits are achieved.

- There may be high social and economic costs.

Contending

Contending is a bilateral strategy that happens when you:

- Have very high concern for your goals and little concern for other people's goals.

- Are not afraid of losing.

- Have high hostility towards the other people.

- Are rigid in your position.

- Are unaware that you can spread the pie, giving each person a part.

- Have the ability to contend.

- Believe that the other people will give in.

Here are some tactics:

- *The principle of liking*: persuade the other party to like you by paying them compliments, doing them favors or demonstrating a positive image.

- *Persuasive argumentation*: convince the other party that they are asking for too much.

- *Offering reciprocal promises*: reciprocal promises have five disadvantages:
  — Cost a lot
  — Can lose effectiveness
  — Involve no voluntary work
  — Very costly to break
  — Very hard to devise

- *Manipulation*: convince the other party, falsely, that what you want them to do is in their best interest.

- *Threats and dirty tricks*: threaten the other party to disclose something they're afraid of.

- *Irrevocable commitments*: mutual disaster may threaten the parties if they break their commitments.

Problem solving

Problem solving aims to find a solution that is acceptable to all conflicting parties. This requires the availability of a true and genuine concern for achieving mutual gains. It also requires the parties to show some flexibility on the proposed solution, although they can stand firm on individual interests. All parties should be open to innovative solutions. Here are the six problem-solving steps:

1. In many cases you discover that the reason for the conflict is a simple misunderstanding that can be easily rectified. Almost half the conflicts I've been required to handle were solved by letting the two parties go away on their own and discuss their differences face-to-face.

2. Discuss what you want and what the other party wants; consider how to make concessions to reach agreement. The cornerstone is that you do not want to be enemies. Prepare your strategy along these lines:
   - Make a list of your interests:
     — concessions you can give away
     — concessions you might trade
     — concessions you cannot make
   - Discuss your problem thoroughly; ask yourself why, why not, and what if?
   - The more thorough your analysis, the quicker your solution will come.

3. Try to locate solutions for both parties to the conflict.

4. Make mutual low-priority concessions using the list you prepared in step 2.

5. Repeat steps 3 and 4 until you come to a mutually acceptable solution.

6. Write down what you have agreed and give a copy to each of the parties involved.

If you are appointed as a mediator, the following strategy can lead you to a fruitful solution:

1. Let each party state their problems, without any interruption from the other parties.

2. Let them talk while you listen, then restate what you have heard to prevent any misunderstanding and to filter out any unnecessary (harsh) statements.

3. If in doubt, ask clarifying questions.

4. Concentrate on the present and talk about the future, never talk about the past.

5. Stick to the topic at hand, do not be diverted onto side issues.

6. Agree whenever you can on clearly stated issues, clarify ambiguous issues.

7. Ask each party to restate their positions and paraphrase what is being discussed, so there is mutual understanding of the issues under discussion.

8. Insure the discussion is concentrated on the topics at hand and does not become sidetracked onto personal issues.

9. Use body language tools effectively; be consistent with your verbal and nonverbal actions to help in transmitting your ideas calmly and nicely to the parties involved.

10. When expressing negative issues, choose your words carefully and never use negative facial expressions.

11. Always show confidence in the two parties, and most importantly in yourself.

12. If the conflict escalates, withdraw temporarily once you have confirmed the place, date and time of the next discussion.

SUMMARY

The disaster management strategy depends on the type of technology. Digital technology causes big losses when it fails, so it requires very careful disaster planning. Fire is the worst and most frequent disaster; it too requires very careful planning.

System reliability is directly linked to cost, and a cost-effectiveness study is needed before deciding to increase it. Reliability can be achieved in the design, the construction and the method of use.

Network configuration and management are important in coping with traffic variations and congestion. Several methods can be used to limit the loss of traffic during congestion.

People have a major impact on how well a system can cope with disaster. Conflict resolution is an important part of disaster management.

REVIEW QUESTIONS

1. Define the following terms: MTBF, MTTR, MDT, A.

2. Give an example from your work experience of a disaster that you identified and prevented by using the techniques in this chapter.

3. Explain the elements you would consider when trying to increase equipment reliability at your workplace.

4. Explain how you would conduct a reliability cost-effectiveness study at your workplace.

5. For each reliability consideration you have identified at your workplace, explain how you might implement it.

6. What are the benefits that you will gain from conducting a cost-effectiveness study at your workplace? How might you measure the benefits?

7. Describe the network configuration at your workplace. What are its interface problems?

8. Outline the network management techniques at your workplace. How might you improve them?

9. Describe the motivation of employees at your workplace. How might you apply problem-solving techniques to improve your employees' motivation?

6

Service Recovery

OBJECTIVES

- Answer questions at the end of the chapter.
- Discuss the different service recovery techniques.
- Analyze service recovery at your workplace.
- Analyze hazards at your workplace.
- Say how to protect your workplace using service recovery techniques.

6.1 THE IMPORTANCE OF STRATEGY

Disaster planning allows companies to absorb the impacts of disaster and prevent telecommunications failures. By planning ahead, companies stand a far better chance of recovering from disaster than they would otherwise. Planning for disaster prevention and recovery does not give a blanket guarantee that a company will always recover from any disaster, but it does provide some measure of recovery assurance; it also helps in preventing avoidable disasters. Finally, companies with tested disaster management plans tend to survive disasters.

Disaster planning and recovery is concerned primarily with the safety of the people in the organization, the equipment they use to provide the service, and the safety of the workplace itself. Disaster planning has two objectives:

- To insure that everything and everybody is protected and safeguarded against any harm or danger that may happen because of a disaster.
- To insure the organization will be able to continue its operations in the intended manner, the predetermined manner in such situations, or will be able to restart again during a certain predetermined period of time.

The disaster planning and recovery strategy concentrates on the smallest individual vital unit (SIVU) in the organization. A SIVU is the smallest important unit without which the organization cannot function as usual. We need to know four things about each SIVU:

- What it does

- What it requires

- What will knock it off

- What happens when it's knocked off:
  — Will the SIVU work normally in a disaster?
  — Will the SIVU adopt an alternative mode in a disaster?
  — Will the SIVU return to its normal mode after a disaster?

When preparing a disaster and recovery plan, it is very important to know how long the organization can function without a particular SIVU. Perhaps it is possible to lengthen this time; two questions spring to mind:

- How can we lengthen the time?

- Which methods are practical?

It usually boils down to economics. Suppose we have a 10 000-line telephone exchange that serves the president's office; its failure rate is 1 per month but each failure means the telephone calls will be disrupted for a period of 20 minutes. And suppose 90% of the calls are international calls, the average international call tariff is US$ 1.00 per minute. On average 20% of the lines are busy all the time.

To reduce the failure rate, the telecommunications authority suggests installing another power supply unit and another mini switch; this will reduce the failure rate to only 1 per year. The investment needed is calculated at US$ 100 000. Will this be profitable?

We have 9000 telephones connected to the international network, with 20% busy at any time. Thus we have an income of

$$9000 \times \frac{20}{100} \times 1 = \text{US\$ 1800 per minute}$$

The cost of interruption is

$$1800 \times 1 \times 20 \times 12 = \text{US\$ 432 000 per year}$$

There will be a gain of US$ 332 000 in the first year then US$ 432 000 in all subsequent years. It seems feasible to go ahead with the project. Now suppose the number of busy lines drops to 4%. Then the annual cost of interruption will be

$$\frac{4}{20} \times 432\,000 = \text{US\$ 86 400}$$

There will be a loss in the first year of US$ 13 600, so maybe the project will not go ahead.

In designing for disaster management, we have to consider two separate planning aspects:

- *Disaster prevention/coping planning*: this means designing the system to prevent a disaster from happening, or to reduce the probability of its occurrence and the level of the risk involved.

- *Disaster recovery/restoration planning*: this means designing the system and including what should be done if a disaster does occur, the steps that need to be taken to insure the organization will not be crippled by the disaster, how it will cope with the disaster properly, and how it will come back to normal operation within the time period specified in the disaster plan.

The combined plan should follow these subplans:

- *Disaster likelihood reduction plan*: this decreases the likelihood that a fault or failure will develop and strike the organization.

- *Disaster prevention plan*: this prevents the likelihood that a fault or failure will occur, develop and impact the organization. The prevention plan aims at minimizing the disaster's impact.

- *Disaster management plans*: these are the detailed tasks or actions designed to counteract the effects of the disaster, and to prevent escalation of failures into disasters.

- *Disaster recovery plan*: this covers the detailed tasks or actions to restore the functioning of the organization, the employees, equipment and workplace hit by the disaster.

6.2 THE DISASTER PREVENTION SCENARIO

There are four elements that constitute a disaster:

- The unusual event causing the disaster
- The impact of the disaster on the organization
- The consequences and effects of the disaster
- The disaster resulting from the impact of the unusual event and its consequences

6.2.1 The disaster communications plan

Promptly communicating an unusual event (a crisis) causing a disaster, to the authorities and employees is a vital role to be played by the disaster management team, since it can mean saving life and reducing losses. There are two cases: disaster due to equipment malfunction and disaster due to other causes.

Disaster due to equipment malfunction

People initially respond to malfunctioning equipment when:

- The alarm indicates an emergency situation:
  — aural alarm
  — visual alarm

- There's no alarm but they sense something is wrong:
  — burning smell
  — unusual sound

Such an emergency situation is never communicated at once to the disaster management team; instead action is delayed until the real cause of the emergency is identified and its impact has been determined. Then it might be too late to take prompt action and the company suffers enormous losses.

Thus it is important that the disaster management team prepare a communications plan that specifies clearly and unambiguously the responsibilities of the operators, technicians and supervisors in relaying the warning signs well before any malfunction develops into a disaster.

Disaster due to fire, flood, etc.

In this case people tend to take the alarm of fire as an indication for a fire drill; they tend to ignore it until somebody in authority tells them the truth and actually forces them to evacuate the place. It is important to take fire drills seriously and make participation mandatory. Clearly marked escape routes should be kept unobstructed, and alternative routes should be clearly marked so that even those who are in the building for the first time can easily follow them to safety. The disaster management team must indicate in their plan the tasks each member of the team will assume in this respect, to insure speedy and quick evacuation.

6.2.2 Handling the responsibilities

Handling a disaster does not begin when the first warning signs become clear. In fact, we must try to insure no event happens that will result in an early warning sign, but we must also try to calculate the occurrence probability of early warning signs before they actually appear. Thus our work must begin when everything is running normally, that is why we have responsibilities before an abnormal event occurs, during its development into a disaster, and after the disaster has been overcome.

Responsibilities before a disaster

The attitude of the people before a disaster is a major factor in the development of the disaster. They usually perform their duties poorly, they fail to detect the early warning signs and even deny their existence. Once the disaster is clear, they exchange accusations about who is responsible for the disaster instead of cooperating to prevent it from disrupting the organization's activities. All in all, they are afraid and angered but they take no action. Thus the logical responsibilities before the occurrence of a disaster should be:

1. Survey the probable and abnormal events that can strike the organization, its people, equipment, products or facilities. These are risks that have to be accounted for in accordance with the occurrence probability of each event.

2. Conduct a vulnerability search to see what is vulnerable to the impact of each of the disastrous events that were identified earlier, and determine the impact on company operations.

3. Conduct a vulnerability analysis to analyze the identified events and rate their effects on the organization according to their severity. If we identify ten events, we might find that only three have really disastrous effects on the organization, thus we have to concentrate on these three.

4. Conduct a vulnerability rectification to find ways of reducing the effects and impact of the identified events, and to establish procedures that make their occurrence a remote probability.

5. Consult an insurance company to determine their charges for coverage.

6. Perform disaster management drills, and make the participation of all employees mandatory.

7. Print the plan in a small book using simple language; distribute it to all company employees.

8. Regularly monitor the operations reports to detect any degradation in quality that might be a future cause of service disruption.

Responsibilities during a disaster

People's attitudes during a disaster play a major part in its development. People usually fail to handle the disaster because they were not trained properly, so they feel upset and worried about the consequences. They have a feeling that they are completely lost, which develops into a situation of total self-collapse. The logical responsibilities during a disaster should be:

1. The disaster management team will perform their duties as per the plan.

2. A member of the team will document the disaster for later evaluation:
   - the sequence of events in the disaster
   - the damage to people, property and equipment

3. A member of the team will record and follow up all notifications to:
   - the emergency services
   - the disaster management team
   - the damage control and restoration team
   - the alternate service site
   - any other notified party, e.g.
     — customers
     — employees
     — managers

Responsibilities after a disaster

People's attitudes after a disaster also play a major part. People are usually shocked, they are unsure of their own capabilities, or even their preparedness to try putting things together to resume service. They also have an urge to make radical changes in the operation and management staff and their responsibilities, so they can be sure that such a disaster will be handled better next time. Yet this might not always be the right thing to do. The logical responsibilities after the disaster should be:

1. Document all fatalities or injuries to personnel.

2. Before any cleaning begins, document the actual damage to infrastructure, buildings, equipment and other property.

3. Document all assets that were damaged beyond repair.

4. Inform the insurance company.

5. Follow up activities leading to the resumption of service.

6. Evaluate the plan itself and the way the disaster management team performed their tasks.

7. Decide on any changes or modifications that should be made to the plan or to the tasks of the team.

8. Train the disaster management team on the proper performance of their duties according to the modified plan.

6.3 RISK MANAGEMENT

Risk means the degree of probability of hazard, loss or danger that an organization is subjected to when confronted by an event that may cause disaster.

Table 6.1 General hazard analysis card

Organization name	Facility name
Date of last analysis	Facility location
Project code	Prepared by
Date	Position
Hazard description	Departments concerned

Hazard parameters			
Severity	Probability	Cost	Action
Minimal	Very unlikely	Minimal	Discard
Nuisance	Unlikely	Nominal	Defer
Marginal	Probable	Significant	Analyze
Critical	Considerable	Extreme	Investigate
Catastrophic	Imminent	Prohibitive	Immediate

6.3.1 Locating and defining hazards

Locating and defining hazards is a two-step process best carried out using three special cards (Tables 6.1, 6.2 and 6.3).

The general hazard analysis card

This gives a general indication of the types of hazard suspected in a certain department and a rough estimate of four parameters concerned with the disaster: the expected severity, its probability of occurrence, the likely cost of the damage and the recommended action.

The detailed hazard card

This is a detail enlarged from the general hazard analysis card. It analyzes an event that seems to have dangerous and disastrous effects on the company, and it includes the following information:

- The task or activity where the disaster is expected to occur
- Details of the hazard involved

Table 6.2 Detailed hazard analysis card

Organization name	Facility name
Date of last analysis	Facility location
Project code	Prepared by
Date	Position
General hazard analysis card number Date	Department concerned
Task/activity	Hazard details
Alternative solutions 1 2 3	Evaluation of each solution 1 2 3
Recommendation	
Recommended solution	Estimated cost
Recommended actions 1 2 3	Recommended tasks By whom When How Where
Situation if action is not taken	
Expected disaster	Estimated cost of loss/damage

Table 6.3 Hazard ranking card

Organization name		Facility name			
Date of last analysis		Facility location			
Project code		Prepared by			
Date		Position			

Probability of hazard occurrence	Effect of the disaster on the organization			
	Very high	High	Medium	Low
Very high	Rank 1 (i) (ii) (iii)	Rank 2.2 (i) (ii) (iii)	Rank 3.3 (i) (ii) (iii)	Rank 4.4 (i) (ii) (iii)
High	Rank 2.1 (i) (ii) (iii)	Rank 3.2 (i) (ii) (iii)	Rank 4.3 (i) (ii) (iii)	Rank 5.2 (i) (ii) (iii)
Medium	Rank 3.1 (i) (ii) (iii)	Rank 4.2 (i) (ii) (iii)	Rank 5.2 (i) (ii) (iii)	Rank 6.2 (i) (ii) (iii)
Low	Rank 4.1 (i) (ii) (iii)	Rank 5.1 (i) (ii) (iii)	Rank 6.1 (i) (ii) (iii)	Rank 7 (i) (ii) (iii)

- The alternative solutions available
- Evaluation of each solution and its estimated cost
- The recommended solution
- The expected disaster if the solution is not implemented
- A cost estimate of any loss or damage during the disaster

The hazard ranking card

This card is used to classify each hazard according to its probability of occurrence and its effect on the organization.

6.3.2   Estimating consequences

The cost of a disaster is one of the measures used in estimating the consequences of not having a disaster management plan. The cost in this case is twofold:

- The cost of fatalities and casualties suffered by the employees and occupants of the company buildings.

- The cost of service interruption, damage to property and equipment, plus the losses suffered by customers due to the service interruption.

6.3.3   Establishing resources

Securing management support is the first step in establishing resources; once the management is satisfied with the disaster management plan, then resources can be secured. Top management will need to understand fully the impact of the disaster on the operations of the company before they agree to commit any resources such as money, time, effort, equipment and staff.

When preparing the disaster management plan, resources must be committed that will reduce the risk of any dangerous or disastrous event occurring, an event that would disrupt the service and/or endanger the employees. If an event has a low risk and a low occurrence probability, it might be unwise to ask for more resources to cover it.

6.4   VULNERABILITY SEARCH, ANALYSIS AND RECTIFICATION

Vulnerability search, analysis and rectification forms the basis of all disaster avoidance and recovery in the telecommunications industry. Not only will it help to protect the organization, but it will also reduce the costs when a disaster does occur. Its essence is to insure the effects of the disaster, if and when it occurs, will be as low as practically possible.

Vulnerability search takes a careful and honest look at all possible vulnerabilities in the organization. Vulnerability analysis determines the levels of disaster inherent in all the vulnerability areas then considers how to reduce them. Vulnerability rectification means taking action to reduce any risks until their levels are practically acceptable.

There are several types of vulnerabilities depending on the risk they pose to the organization. Table 6.4 details the different types along with the proposed search, analysis and rectification work.

6.4.1   System vulnerability

The system vulnerability search, analysis, and rectification work aims to produce a system that does not suffer service interruptions and the accompanying financial losses. Here are the main items to be considered:

Table 6.4 Vulnerability search, analysis and rectification

Organization name		Facility name	
Date of last analysis		Facility location	
Project code		Prepared by	
Date		Position	
Vulnerability			
Type	Search	Analysis	Rectification
Poses extreme risk (such as improper storage of paints)	Found at once (check the stores)	Easy to analyze (storage regulations)	Work should start immediately to rectify
Poses unnecessary risk (such as blocked fire escape routes)	Easily found (check the firefighting preparedness)	Easy to analyze (fire regulations not properly followed)	Work should start immediately to rectify
Poses apparent risk (such as loss of market share)	Needs properly organized search work to find	Needs careful analysis of the business environment	Work should start immediately to rectify
Hidden risk (such as network management probabilities)	Needs careful search work for concrete data	Needs careful analysis of traffic data	Work should start immediately to rectify
Poses minor and remote risk (such as a fault in the second spare generator)	Needs proper search for its probability of occurrence	Needs careful analysis of the possible impact and the loss involved	Work should start to consider the appropriate action
Poses no risk to the organization	Search to gather data	Analyze to ascertain it is not a vulnerability	No action taken if it is not a vulnerability

- Insure the employees are safe at all times.
- Insure the workplace is safe at all times.
- Insure the service will continue normally without any interruption.
- If there is an unexpected event, insure immediate action will be taken to prevent service interruption.
- If there is a service interruption, implement any service plans immediately.

The main activities can be grouped as follows:

- Search for vulnerabilities that may cause death or injury to employees.
- Search for vulnerabilities in the organization's buildings; check the structural stability of the buildings, the stores and other contents of the buildings to insure they pose no risk.
- Check the data collection system in the organization; insure all data items are current then check the probabilities of data vulnerability, determine the source of such vulnerability and the appropriate rectification.
- Check the elements that might cause a service interruption if an unusual event occurs. What contingency plans are required? Are they sufficient to take care of the situation? If not, what should be done, when and how?
- Check any conditions that might happen over time, perhaps causing the building or the system to fail. Check the contingency plans cater for these time-dependent situations.
- Check the contingency plans have addressed all the cases identified as system vulnerabilities.

6.4.2 Personnel vulnerability

It is very difficult to analyze the human causes of disasters since we know very little about the way humans will behave when a certain disaster occurs, and how they will react. Nevertheless, when analyzing disasters, one should always consider any human factors; most of them fit into five categories:

- *Operating errors*: such as maloperation or errors in procedures; we can also add external events like electric shock or damaging a cable.
- *Erroneous errors*: such as programming errors.
- *Maintenance errors*: such as lack of maintenance procedures, improper maintenance performance, malpositioning and negligence.
- *Test errors*: such as improper meter reading and calibration errors.

- *Design errors*: such as the lack of quick and easy access to the different parts of the equipment.

Experience indicates that considering all these factors, it is a certain small group of operators or technicians who commit the majority of such errors, causing disasters to the organization. Transferring these technicians to a different section and/or giving them training on disaster prevention techniques does not reduce the rate at which they cause disasters. These people have what may be called disaster proneness. This means that some people have a personality trait which predisposes them to accidents. It probably depends on their age, experience, hazard exposure, work environment and on their own unique personality.

It might be useful to identify and analyze the people causing repeated accidents then try to transfer them to other parts of the organization where they can cause minimum harm.

6.5 CONTINGENCY MANAGEMENT

Contingency plans are usually developed after a contingency strategy has been decided but it is advisable to treat contingency planning as a main part of formulating the contingency strategy, because a new strategy might develop from examining new contingencies, allowing the response to be more flexible.

Sometimes people are not happy with planning for a disaster, on the assumption that everything is running normally and nothing wrong is anticipated in the near future. This is a very dangerous attitude. Even if things are in good shape now, we can never be sure of the future, as Murphy's law states. A simple contingency plan is better than nothing, since it can have great benefits if and when the unpredictable occurs.

Contingency planning begins by identifying the most dangerous event; this is predicted from the worst-case scenario that is developed. A trigger point must be identified which indicates when the dangerous event has developed sufficient impact, the probability of a disaster is high, and the contingency plan can be implemented.

Estimating the required funding for the contingency plan is a very important factor in the success of the planning process, since without funding nothing can be done. An inflated budget will make the management hesitant in approving the funds for the plan. Thus it is advisable to make the funding requirements realistic, readily available and correlated to the anticipated dangers.

6.5.1 Control and minimize disaster

The main task of the disaster management or contingency management team is to control and minimize the impact of the unusual event that caused the disaster. Thus the aim of the disaster management efforts is to predict and prevent the occurrence of this event, but if it does actually occur then the plan must handle the harmful effects and insure service continuity.

6.5.2 Rescue and safeguard people

The first task of the disaster management or contingency management team, in case of fire or any life-threatening event, is to rescue and safeguard the employees or occupants of the company's premises. A loss of life can be very expensive to the company. This includes containment of the fire, safe passage through the escape routes and any medical help given to the injured.

6.5.3 Secure safe rehabilitation

The main task of the disaster management or contingency management team, during a non-life-threatening disaster, would be to secure rehabilitation of the service. Yet a contingency plan can be thought of that gives certain teams different tasks to be performed simultaneously. This means that while group 1 is trying to control and contain the disaster, group 2 is helping to rescue the trapped employees and escorting them outside the building that caught fire. Group 3 would be trying to establish the alternative communications facility so the service can be restored as quickly as possible. Yet in this case the cost of the plan will increase. A cost-benefit analysis will indicate which plan to adopt.

6.6 THE DISASTER-HANDLING SCENARIO

Analyzing the disaster ground

This must not necessarily be taken as a geographical task, but it should be looked at in its general form of analyzing the problems that appear to control the disaster operation theater, such as the points of power for each party to the disaster, and the types of problem that may face the disaster-handling team.

Analyzing the general situation

From the viewpoint of those who originated or generated it, the disaster is not a target in itself but it is a means to achieve a certain goal. So analyzing the general situation makes the disaster-handling team discover the real cause of the crisis and the ultimate goal of the resulting disaster.

The analysis begins by assessing the crisis situation and the current stage of the crisis: defining what powers are facing us, the anticipated dangers due to any delay in taking action, the amount of anticipated losses due to this situation. Then we may find it better to discuss three policies:

- *Long-term confrontation*: this extends the time frame of the crisis so the opposing parties lose their power; it is sometimes used during industrial action. The management knows the workers cannot withstand living without pay for more than a specific number of weeks, so if the anticipated production losses during the strike are less than the workers' pay claim, the management might adopt this strategy.

- *Quick confrontation*: this gains quick control over the crisis. The disaster-handling team resorts to quick confrontation when the effects of the crisis are detrimental to the health of the organization, or if they will have major security, political or social implications.

- *Calming*: the disaster-handling team may use calming when the crisis has achieved all or most of its goals and the situation needs to be calmed to cope with a new environment arising out of the crisis.

Task analysis

The tasks of the disaster-handling team fall into several different categories:

- *Simultaneous tasks* should be executed at the same time and in complete synchronism, otherwise major problems may occur.

- *Successive tasks* should be executed one after the other to achieve the expected goal.

- *Separate tasks* are necessary to diffuse the crisis.

Resource analysis

The team will analyze all the resources they have and match them with the tasks they are required to perform. The analysis must include a discussion of the available chances, the available staff, the qualifications of the staff and the crisis training they have received.

Consider the possibility of isolating the source of the crisis, which may develop into a disaster, e.g. a fire in a power station engine room. How can the room be isolated to stop the spread of fire to the second engine room? Do we need special kit for that task?

Modern crisis prevention elements — up-to-date records, circuit diagrams, test equipment, etc. — are important assets in preparing for a disaster.

6.6.1 Basic scenario planning

Each disaster has its own special parameters, so it is hard to work from a few basic scenarios; they will have to be devised to suit the circumstances.

A scenario is a tool to achieve a certain goal through the distribution of specific tasks, with specific timing, to each and every member of the disaster management team. These integrated tasks, when completed, will achieve the successful termination of the disaster.

Several alternative scenarios should be drawn up; this gives the disaster-handling team the necessary freedom to handle changing situations when confronting the disaster. Here are the elements that should be written into each scenario.

The form of intervention

Explain the form of intervention for each type of disaster, and furnish several alternatives that cover most of the anticipated disasters. When fire erupts in a building, the first task for the team members is to try to put it out while someone contacts the fire brigade.

Limits of responsibility

Clearly and without ambiguity, determine the limits of responsibility for the different members of the disaster management team. Then they can work in harmony to confront the disaster and restore the service as quickly as possible.

6.6.2 The domino effect during disasters

A domino effect occurs when the initial disastrous event is responsible for creating a series of other disastrous events — a chain reaction. Discuss and carefully analyze the nature of the disastrous event; this will eliminate the possibility of a domino effect.

A domino effect usually leads to greater losses. Since disaster management aims to minimize losses, it is worthwhile noting any consequences of a domino effect.

6.6.3 Best-case and worst-case scenarios

The best-case scenario assumes everything will go according to plan; the worst-case scenario assumes everything will go wrong. The moderate-case scenario, or the most likely scenario, assumes things will go right most of the time but they will sometimes go wrong. In the most likely scenario we take note of this fact and add a factor to cater for the best and worst cases. Here is an example.

Execution time within a PERT program

The program evaluation and review technique (PERT) is a powerful tool to calculate the most efficient way to implement a project (see pages 36–40). There are two versions:

- *PERT/time* calculates the shortest time to implement the project.

- *PERT/cost* calculates the lowest cost to implement the project.

Time is important because a project completed in 12 weeks instead of 16 can begin generating profits 4 weeks earlier.

To perform a PERT analysis, one must decide on the implementation period of each activity within the project; this is very difficult since sometimes an activity can be performed in a short time and sometimes it takes much longer. The way to solve this problem is to assign each activity three values:

- The shortest time T_s, when things are very good

- The longest time T_l, when things are extremely bad

- The most likely time T_n, when things are normal

We consider that things are normal in 66% of cases, very good in 17% of cases, and rather bad in 17% of cases. Thus, to determine the most probable time, we use the following equation:

$$\text{Most probable time} = \tfrac{1}{6}(4T_n + T_s + T_l)$$

6.7 A GENERIC DISASTER MANAGEMENT PROPOSAL

A generic disaster management proposal has four main steps:

1. Establish the main strategies, indicating the main events that can endanger the operations of the organization.

2. Develop the necessary contingency plans to cater for the dangerous events predicted in the strategy. This will be done for the worst-case scenario of the strategy.

3. Calculate the financial implications and the expected time of interruption for each dangerous event; secure management approval for the necessary funds.

4. Select and train the disaster management team; proper and efficient implementation of the plan is vital to its success.

6.7.1 A model program

- From the company organization chart, we can identify all work units, their function and their managers.

- Identify the equipment installed at the company. If you are not familiar with any of the equipment or systems, have a casual meeting with the responsible person and obtain clarification.

- Identify areas of risk or hazards and record them on the hazard control sheet.

- Identify telecommunications system facilities connecting the company to the external world, and check if there are duplicate circuits.

- Research the company's disaster recovery file and see how it faced previous disasters.

- Contact the insurance company that deals with your organization and make a note of any concerns it may have about your business.

6.7.2 Data collection and analysis

They are many reasons to collect data about business functions and resource requirements. A questionnaire is a good way to collect data.

- To calculate the loss potential used in justifying the disaster-planning expenditure to the management.

- To help choose a cost-effective disaster management application.

- To keep records up to date, readily available and properly stored.

- To set realistic disaster objectives and to devise appropriate recovery strategies.

6.7.3 Preparing a questionnaire

General rules

Form of the questionnaire

- The purpose must be stated in the questionnaire itself. The respondents have to be told the purpose. To a large extent, the quantity and quality of the replies will depend on how well the researchers have managed to persuade the respondents about the usefulness of the survey.

- The respondents must be able to answer the questions easily and quickly. The questions must therefore be clear and precise; avoid ambiguity. If possible, use precoded replies (boxes to be checked) so the respondents need not write a sentence. But also provide space for people who do want to write something.

- The questionnaire must be presented attractively. You may ask the respondent to identify themself by name, but always preserve their anonymity. Include any completion instructions, return address and deadline date.

Substance of the questionnaire

Before designing the questions, write down the types of information you wish to gather:

- General information about the subject
- Existing knowledge within the target population
- Major difficulties that may be expected
- Differences of view within the target population
- Motivation towards the survey

Presentation of the questionnaire

- Write specific questions for each domain and select questions that are easy to answer, easy to process, and elicit the greatest amount of useful information.
- If several questions on the same point are required to check the accuracy of people's answers, make sure they are scattered throughout the questionnaire.
- Try out the questionnaire on a few members of the target group. Check the questions can be understood and they generate useful responses. This is the validation process.

Processing information

- Tabulate the results properly; this will help to summarize all the replies in one document and will make them easier to analyze.
- Interpret the results using common sense. Look at what people have actually written not what you think they might have written.
- Prepare a report for inclusion in the final analysis.

Specific guidelines

Conditions to be taken into account

- Phrase questions in the present tense.
- Deal with specific cases that can be accepted or rejected.
- Keep the questions short, clear and easy to read.
- Restrict each question to one idea.
- Do not ask about things that are ambiguous, invalid or incomplete.
- Avoid questions that elicit a unanimous yes or no.

Selecting the main cases

- Choose some general topics concerning the telecommunications breakdown, fire, flood, earthquake and lightning.

- Ask some judges to arrange these cases into five groups according to the target group's general trend of belief towards disaster management in each area:
  — Group 1 deals with those cases which the judges believe to be very supportive of the trend.
  — Group 2 deals with those cases which the judges believe to be supportive of the trend.
  — Group 3 deals with those cases which the judges believe to be neutral.
  — Group 4 deals with those cases which the judges believe to be slightly against the trend.
  — Group 5 deals with those cases which the judges believe to be highly against the trend.

- The topics that the judges find to be highly supportive of the general trend, they should be chosen as fundamental cases since they truly represent the target population's perception of the real trend of disaster management in their telecommunications facility.

- The topics where the judges slightly differ in their views should be treated as minor cases.

- The topics where the judges differ in their views should be discarded since they do not represent the trend of thought in the target population.

Sample questionnaire

Table 6.5 considers service interruptions to high-frequency links and classifies them according to their symptoms; this is the only way we can detect an unusual event. Each symptom is identified by four parameters:

- The position in the network or circuit where the fault occurred.

- The frequency of occurrence of this fault; if it occurs frequently then more attention must be given to its rectification.

- How the fault affects the network or circuit; if it has grave effects then immediate attention should be given to it.

- The period of interruption gives an indication of the losses encountered; the greater the loss, the more urgent the intervention.

We also classify the cause of the fault according to equipment failure, lightning, flood, fire or any other explanation. An important element in Table 6.5 is the fault reporting—when the fault was reported and

Table 6.5 Questionnaire for faults causing service interruption

Symptoms					
	Position	Frequency	Effects	Period of interruption	Remarks
Consistent					
Intermittent					
Breakdown					
Degraded performance					
Other (specify)					
Cause of fault					
Equipment failure					
Lightning					
Flood					
Fire					
Bad handling					
Reliability					
Other					

(*continued overleaf*)

Table 6.5 *(Continued)*

Fault reporting					
	Reporting officer	Method of reporting	Time elapsed	Accuracy of reporting	Remarks
Usually					
Occasionally					
Seldom					
Never					

Place of repair					
	On site	Unit workshop	Main shop	Manufacturer	Remarks
Usually					
Occasionally					
Seldom					
Never					

Fault due to					
	Component	Connector	Solder	Bad handling	Other
Usually					
Occasionally					

Table 6.5 (*Continued*)

Fault due to					
	Component	Connector	Solder	Bad handling	Other
Seldom					
Never					

Failed item					
	Specify	Specify	Specify	Specify	Remarks
Usually					
Occasionally					
Seldom					
Never					

Recommended action for frequent faults					
Action					
Funds needed	$		Staffing		Time	

Recommended action for disastrous faults					
Action					
Funds allocated	$		Staff provided		Time to implement	

how — since this is vital in calculating the actual time taken to repair the fault.

Completion instructions
Symptoms and causes of fault

- *Position*: the position in the system or the network where the fault occurred.

- *Frequency*: how often the fault occurs (high, moderate, low).

- *Effects*: how the system is affected (high, moderate, low).

- *Interruption*: how long the system is interrupted, classified according to its operating parameters.

Fault reporting

- *Reporting officer*: who reported the fault.

- *Reporting method*: telephone, telex, order wire, etc.

- *Elapsed time*: how many minutes from the moment of occurrence to the moment of reporting,

- *Reporting accuracy*: if a fault is incorrectly reported, it may take longer to repair.

Detailed analysis of failed components

We decide to do an in-depth analysis into the causes of failure for items which have a high failure rate.

Recommended action for some types of fault

We differentiate between frequent faults that might not be disastrous, and those faults which can cause a lot of harm to the organization.

6.7.4   Disaster management checklists

1. Select a disaster management team.

2. Collect data about the company's business functions and the end user's business.

3. Perform vulnerability search, analysis and rectification.

4. Grade the vulnerabilities according to their effects on the company.

5. Develop a recovery plan to cater for the most serious disaster (system, network and user plans).

6. Obtain management approval for the plan.

7. Obtain the necessary resources.

8. Train the disaster management team to carry out the plan.

9. Train the employees on their disaster management responsibilities.

10. Conduct disaster awareness drills to insure the correct performance of the team and employees during a simulated disaster situation.

6.7.5 Presentation of the plan

The management have given the go-ahead to prepare a disaster management (recovery) plan. The team responsible for the plan have studied and prepared the initial phases of the plan. These are:

1. Assess the vulnerabilities within the organization.

2. Analyze the organization to assess its crisis management capabilities.

3. Assess the organization's preparedness for coping with a disaster.

4. Conduct a risk assessment study.

5. Prepare a disaster prevention plan.

6. Prepare a system recovery plan.

Now present this plan to the management to secure funding and begin implementation. Follow these simple rules.

Before the presentation

1. Insure the plan addresses present-day requirements and also looks into the future.

2. Do not make the plan too detailed; a good plan should be no more than three to four pages long, so busy managers have time to read it.

3. Prepare answers to all the expected questions; find out the FAQs and produce an answer for each.

4. Give an optimistic view of the disaster management situation in the organization.

During the presentation

1. Begin by exposing the most dangerous risk that the organization is facing; once you have the management's attention, you can proceed with the rest of your presentation.

2. Next you present the disaster avoidance capabilities already existing at the organization, how much they cost, and the cost of elevating them. Give good reasons for your costings.

3. Now you are ready to present the system and network recovery strategies along with the costs involved.

4. Explain the required insurance coverage and its costs; as the company becomes more capable of handling a disaster, the cost of insurance will decrease.

5. Last comes the training plan for teaching the company employees how to prevent and cope with a disaster.

After the presentation

- You must be able to answer all the questions you are asked. If some questions need figures that you do not have, tell the manager concerned that you can give them the answer later on; specify how long it will take, say 24 hours.

- Once the disaster management proposal has been approved, along with the appropriate funds, perform the tasks needed to implement the plan.

The implementation plan

1. Select the disaster management implementation team.

2. Distribute the tasks between the team members.

3. Amend the existing disaster management capabilities or purchase and install new ones.

4. Develop the system disaster recovery plan to cater for the added capabilities, and prepare procedures to be followed in case of a system disaster.

5. Develop the network disaster recovery plan to cater for the added capabilities, and prepare procedures to be followed in case of a network disaster.

6. Develop the user disaster recovery plan to cater for the added capabilities, and prepare procedures to be followed in case of a user facility disaster.

7. Implement a full-scale training program for the employees on the developed disaster management plan.

6.7.6 Disaster recovery training

Training development guidelines

In 1979 the International Telecommunications Union, a specialized agency of the United Nations, prepared its training development guidelines (TDG) manual to be used in preparing telecommunications training courses. I believe it is one of the best methods available for preparing training programs.

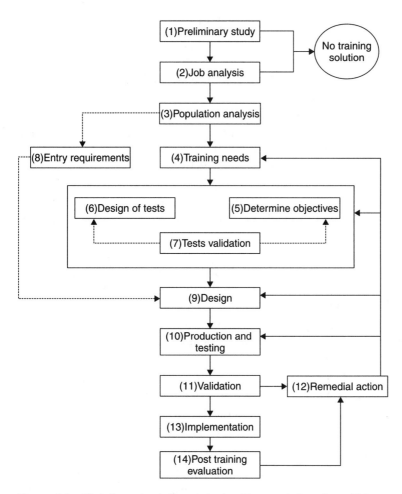

Figure 6.1   TDG flowchart (Reprinted, with permission, from ITU)

The process, termed CODEVTEL (course development for telecommunications), is composed of 14 steps (Figure 6.1):

- *Preliminary study*: to determine that training can solve our problem. Training cannot solve some problems. We have to insure our training budget will have a reasonable ROI.

- *Job analysis*: the job is analyzed to see if it is performed in the best possible way. Does the method have a potential risk that can be reduced by modification?

- *Population analysis*: what sort of people are we going to train? To achieve maximum benefit, different training courses must be prepared for people at different levels of technical expertise and administrative ability.

- *Training needs and job aids*: we decide on the training needs of each population group that require training, and the training aids needed for them to acquire a full understanding.

- *Determining the course objectives*: this enables the course developer to select appropriate training material.

- *Design of mastery tests*: the mastery tests are designed before the training material is written. Then we can be sure the training material will be oriented nicely around the subject, enabling the trainee to answer the success criteria more effectively.

- *Validation of tests and objectives*: this is a very important step since we compare the objectives with the success criteria to insure the trainee who passes will probably be able to cope with the disaster situation they have been trained for.

- *Determining the entry requirements*: this looks at the prerequisites for acceptance into training. If some groups are lacking the basic requirements, they can be given preliminary training before passing on to the main course.

- *Design*: the training course material is now prepared so it satisfies the objectives and success criteria; it is adjusted to suit the level of the course's target population.

- *Production and development testing*: the course is produced then tested by trying it on a few trainees from the target population.

- *Validation*: the test results from the trial group are checked against the test results from the control group, which was not given the course.

- *Remedial action*: this is taken after receiving appropriate feedback.

- *Implementation*: the course is presented to the target groups then feedback from course presenters and trainees is monitored carefully.

- *Posttraining evaluation*: to evaluate how the course is going and the reaction of the trainees' employers regarding how they cope with the disasters they have been trained for.

Advantages of TDG

- It insures the training is adapted to local conditions.
- It insures that relevant information is provided to the trainee.
- It simplifies the adaptation of training material to different population groups and improves communication.
- It enables a complete evaluation of all aspects of the training process.
- By setting the training objective at the start of the process, it helps us to evaluate the learning level.
- It is applicable to any training development project.

TDG forms

Several forms are used with TDG to simplify the job of course developers. The forms on pages 202–209 were developed from items already available, but they include some modifications I have added over the years.

Course description form

The course description form (Table 6.6) is used to describe the course and to identify the general requirements:

- *Tasks to be performed*: they are stated here so it is easy to correlate them with the main points in the course.

- *Entry requirements*: needed skills, knowledge and attitude. They are stated at the end of the form. A trainee must have certain prerequisites so they can pursue the course successfully.

Course content form

The course content form (Table 6.7) is used to describe the course content and to identify the general requirements:

- The number of modules, the objective and duration of each module.

- Training techniques such as lectures, demonstrations, group discussions, tutorials, case studies, lab exercises, program study, independent study, field visits, on-the-job training.

Course timetable form

The course timetable form (Table 6.8) is used to describe the course timetable and to identify the general requirements:

- *Trainee reference*: books, handouts, circuit diagrams, layouts or other references.

- *Instructor reference*: the instructor guide and any other relevant reference.

- *Audiovisual aids*: overhead projector, slide projector, data show, video or any other audiovisual equipment.

- *Equipment*: indicate the equipment used for demonstrations and laboratory experiments.

Module/lesson plan form

The module/lesson plan form (Table 6.9) is used to describe the module/lesson plan and to identify the general requirements:

- *Behavioral objectives*: what the trainee will be able to do, under what circumstances, and to what standard.

Table 6.6 Course description form

Course Description Form	
Course title	Issue date
Course code	Issued by
Duration days	Position
Purpose of the course 1 2 3	
Tasks to perform after successfully completing the course 1 2 3	
Main points to be covered in the course 1 2 3	
Entry requirements 1 2 3	

Table 6.7 Course content form

Course Content Form				
Course title		Issue date		
Course code		Issued by		
Page no		Position		
No	List of modules	Objective	Duration	Training technique
1				
2				
3				
4				
5				
6				
7				
8				
9				
10				
11				
12				

- *Main topics covered*: specify main learning units (subject headings, learning tasks).
- *Key points*: the core of the topic.

Material and equipment form

The material and equipment form (Table 6.10) is used to describe the materials and equipment needed to present the course, and to identify the

Table 6.8 Course timetable form

Course Timetable								
Course title					Issue date			
Course code					Issued by			
Page no					Position			
Day	Period	Mod. no	Mod. title	Trainee ref.	Instructor ref	A/V aids	Equipment	Remarks
1	1							
	2							
	3							
	4							
2	1							
	2							
	3							
	4							
3	1							
	2							
	3							
	4							

general requirements, e.g. the types and quantities of materials to be used in training and their reference numbers. These materials should be ready before the course is presented.

End-of-module mastery test form

The end-of-module mastery test form (Table 6.11) is used to describe the end-of-module test and to identify the general requirements:

- Test conditions that must be satisfied before the test can be conducted.

Table 6.9 Module/lesson plan form

Module/Lesson Plan Form				
Course title		Issue date		
Course code		Issued by		
Subject		Position		
Lesson no		Page no		
Behavioral objectives				
Time	Main topics covered	Key points	References	Remarks

Table 6.10 Material and equipment form

Material and Equipment Form				
Course title		Issue date		
Course code		Issued by		
Subject		Position		
Lesson no		Page no		
Size of trainee group Type of training (self-paced or group paced)				
No.	Quantity	Specs of material/ equipment	Reference	Remarks
1				
2				
3				
4				
5				
6				
7				
8				
9				
10				
11				
12				

Table 6.11   End-of-module mastery test form

End-of-Module Mastery Test Form				
Course title		Issue date		
Course code		Issued by		
Module title		Position		
Test conditions 1 2 3 Success criteria 1 2 3				
Outline of mastery test				
No	Main activities to be tested	Key success points	Reference	Remarks
1				
2				
3				
4				
5				

- The success criteria that the trainee is supposed to verify.
- Outline of the mastery test to clarify the main points.

Disastrous event analysis form

The disatrous event analysis form (Table 6.12) is used for analyzing disastrous events. The real cause of the event is identified and practical remedial actions can be suggested.

Table 6.12 Disastrous event analysis form

Disastrous Event Analysis Form	
Date of the event	Issue date
Time	Issued by
Duration	Position
Location	
Consequences	
1	
2	
3	
Persons involved	
Name	Position
Name	Position
Name	Position
Name	Position
What happened	
1	
2	
3	
How it was handled	
1	
2	
3	

Table 6.12 (*Continued*)

How it should have been handled		
1		
2		
3		

What went wrong		
1		
2		
3		

Why things went wrong	Yes	No
Employees never learned the task performance properly		
Employees forgot how to perform the task properly		
Employees did not understand the importance of performing the task properly		
Inadequate supervision		
Inadequate standards for performing the task		
Inadequate equipment or tools		
Poor organization		
Inadequate working environment		
Poorly defined tasks		
Others (specify)		

Suggested remedial actions		
1		
2		
3		

6.7.7 Testing the disaster management program

Whenever a large-scale disaster management plan is proposed for implementation, it's strongly recommended to test it on a small scale. There are several reasons why this test is valuable:

- To reveal any flaws in the plan.
- To receive feedback on any problems while implementing the plan.
- To check the system response and the network response to the suggested restoration and recovery procedures.
- To correct any interface problems between existing capabilities and any proposed capabilities.
- To train the disaster management team, on the job, to perform their specified tasks within the plan.

Maintaining political broadcasts

As a young engineer in the TV and broadcasting service, I was responsible with a colleague for the operation and maintenance of two high-power transmitting stations. Since the programs broadcast from these stations were politically oriented, it was the management's viewpoint that any interruption in the service, for more than a specified period of time, would be treated as sabotage. All the security measures were implemented:

- Entry permits
- Unannounced security checks
- Guarding the area 24 hours a day
- Firefighting equipment

So the only dangerous event that could happen was a technical failure. Since we had no redundancy at the station, we were forced to predict the fault and clear it in the shortest possible time; see Table 7.3. The disaster management plan was simple; here are its main points:

1. Prepare a fault tree for each part of the station.
2. Identify the reliability standard of the components that are expected to fail and change them before they fail.
3. Learn the station circuit diagram by heart, so you have it clear in your mind when you have to tackle a fault.
4. Check the accuracy of your memory by creating fictitious failures, when the station is off air, then perform faultfinding and record the time to clear the fault.
5. Conduct contests where one engineer deliberately creates a fault in the station, and another engineer tries to clear it within the time limit.

Disaster management drills

The disaster management plan must be tested at intervals to refresh everyone's memory. But these tests do cost money, so their frequency should be determined in consultation with the management. Explain to management that testing is vital to keep the plan up to date with any changes in the organization, the system or the network. It is recommended to test the plan at least every six months.

6.8 SELECTED APPLICATIONS

6.8.1 End user recovery planning

- Identify the business areas that will be affected by the interruption of service that might happen in the company supplying the service.
- Determine what alternative services could be provided in case of interruption.
- Estimate the cost of the alternative service to be provided.
- Prepare a mechanism for notifying the end user with the interruption, and to inform the alternative service provider to begin delivering the service.
- Prepare a detailed plan to be presented to management and the end user, for their approval.
- Train the end user on the approved disaster management plan.

6.8.2 Systems recovery planning

- Identify the system areas that will be affected by the interruption of service.
- Determine what alternative service options could be considered in case of system interruption.
- Check the disaster management resources available for the system at the company.
- Estimate the cost, advantages and disadvantages of each alternative service option to be provided.
- Prepare a communications plan for notification and handling of the disaster.
- Prepare a mechanism for notifying the end user with the systems interruption, and to inform the alternative service provider (if any) to begin delivering the service.
- Prepare a detailed disaster management plan to be presented to management, for their approval.

- Train the disaster management team and company personnel on the approved disaster management plan.

6.8.3 Network recovery planning

- Identify the network routes that will be affected by the interruption of service.

- Determine what alternative route options could be considered in case of interruption to the main route.

- Check the disaster management resources available for the network at the company.

- Estimate the cost, advantages and disadvantages of each alternative route option.

- Prepare a communications plan for notification and handling of the disaster.

- Prepare a mechanism for notifying the end user with the systems interruption, and explain any traffic limitations of the alternative route.

- Prepare a detailed disaster management plan to be presented to management, for their approval.

- Train the disaster management team and company personnel on the approved disaster management plan.

SUMMARY

Planning for disaster helps a company to absorb its impacts and to prevent failures of the telecommunications system. Planning ahead gives the company a far better chance of recovering from the disaster. It does not give a blanket guarantee that the company will always recover from any disaster, but it does provide some measure of recovery assurance and it helps in preventing avoidable disasters. Companies with tested disaster management plans tend to survive disasters.

The disaster communication plan is important for promptly communicating an event causing a disaster to the authorities and employees. It has a vital role to the disaster management team, since it can mean saving life and reducing losses. Each member of the disaster management team should be fully aware of their responsibilities and the limits of their authority before, during and after the disaster.

Risk means the probability of hazard, loss or danger that an organization is subjected to when confronted by an event that may cause disaster. Risk management is how to handle the situation effectively and efficiently. One of the most important elements in preparing a disaster management plan is to search for, analyze and rectify any vulnerabilities that might exist in the organization.

Sometimes people are not happy with planning for a disaster, on the assumption that everything is running normally and nothing wrong is anticipated in the near future. This is a very dangerous attitude. Contingency planning begins by identifying the most dangerous event. A trigger point must be identified; this indicates when the dangerous event has developed sufficient impact, the probability of a disaster is high, and the contingency plan can be implemented. Contingency plans are prepared to control and minimize disaster, to rescue and safeguard people, and to secure safe rehabilitation.

Disaster recovery training is the final step in the disaster preparations. Training involves preliminary study, job analysis, population analysis, training needs and job aids, determining the course objectives, design of mastery tests, validation of tests and objectives, determining the entry requirements, design, production and development testing, validation, remedial action, implementation, and posttraining evaluation.

REVIEW QUESTIONS

1. Define the following terms: communication plan, hazard, risk, vulnerability, contingency, domino effect.

2. Give some examples from your work experience of how a hazard has been detected and analyzed.

3. Explain how you would think methodically about searching for vulnerabilities at your workplace.

4. What are the types of vulnerability that exist at your workplace? Give three examples.

5. Explain how you would analyze the detected vulnerabilities and attempt to rectify them at your workplace.

6. When preparing a crisis-handling scenario, what are the duties and responsibilities of the disaster management team?

7. Explain how you would prepare a disaster management plan for the company you are working at.

8. Explain how you would present the disaster management plan to the company's management.

9. Having obtained approval for your disaster management plan, how would you prepare the training plan to back it up?

10. How would you estimate the total cost of a disaster management plan, and what are the different subcosts involved in this estimate?

7

Options for Disaster Recovery

OBJECTIVES

- Answer questions at the end of the chapter.
- Discuss basic reasons for telecommunications disasters at your workplace.
- Identify the factors affecting telecommunications disasters at your workplace.
- Explain how to handle a telecommunications disaster at your workplace.
- Describe some disasters that happened at your workplace.
- Explain how to evaluate the cost of a disaster at your workplace.
- Consider any security and political issues of disasters at your workplace.

7.1 BASIC REASONS FOR DISASTER

7.1.1 The early warning signs

Figure 2.1 illustrates how the O&M engineer should tackle the problem before it develops into a disaster, by detecting and properly handling the early warning signs. It also indicates the grave consequences of ignoring or misinterpreting the early warning signs. Detecting the early warning signs means observing any slight change in the normal operating parameters of the system under consideration.

7.1.2 Types of telecommunications disaster

Table 7.1 shows the major disasters that affected telecommunications services in the United States during the period 1975–91. The majority of disasters were attributed to natural causes such as hurricanes, floods, earthquakes or fires. The second major cause was software problems.

Thus, no major disaster was reported due to a purely technical problem in the telecommunications equipment itself, and this can be attributed to the use of redundancy techniques in all areas of the telecommunications network, so an alternative route is put into service once the main route

Table 7.1 Major US telecommunications disasters 1975–91

Year	Type of disaster	Area affected	Consequences
1975	Fire	Second Avenue Central Office, New York	Disruption of service, 170 000 lines affected
1987	Fire	Bushwick Central Office, New York	Disruption of service, 40 000 lines affected
1988	Fire	Central Office, Hinsdale, Illinois	Disruption of service to about 0.5 million customers
1989	Hurricane	Telecommunications facilities at Charleston, South Carolina	Loss of electrical power disrupted telecoms services
1989	Earthquake	Telecommunications facilities at San Francisco, California	Loss of electrical power and ripped cable connections disrupted telecoms services
1989	Tornadoes Floods Hurricane	Several parts of the United States	Loss of communications services
1990	Software glitch	AT&T	Many customers lost communications services
1990	Cable cut	Illinois Bell	Many financial houses and radar locations lost communications services; local airports were disrupted
1991	Computer system software glitch	Washington DC and Los Angeles	About 9 million lines were impaired, affecting local and cellular services

becomes faulty. This idea cannot be applied if the problem is in the software that controls the network, because the software does not fail in a disaster, it just works irrationally. Finding this computer glitch might not be easy every time it occurs. Minor disasters no doubt occurred but they were not reported because their effects were minimal: small fires, power loss, cable cuts, vandalism and computer viruses.

The disasters due to fire can be eliminated by a proper firefighting plan (Chapter 9).

7.1.3 Stages in a disaster (Figure 7.1)

• *Durable condition*: this is the normal (steady-state) condition, where all is well and all operations are running smoothly.

• *Stable condition*: this is a quasi-normal condition; the operations are running smoothly but the QoS figure is at its limit. Suppose the distortion of the output wave is meant to be not more than 1% and does not exceed 0.5% in normal operation. If the distortion content reaches 0.5% then it is still within the limit but something is not okay.

• *Early warning signs*: here we get a warning sign; perhaps the pilot lamp indicates the bit error rate (BER) is high, or perhaps the heat detector indicates a rise in temperature, maybe from a fire.

• *Problem occurs*: the problem is now clear; if it is not handled fast, it will develop.

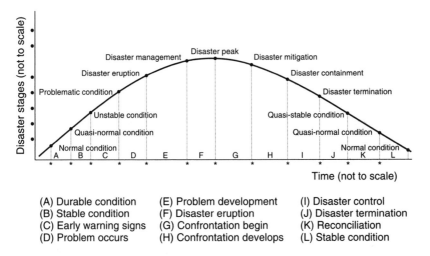

(A) Durable condition	(E) Problem development	(I) Disaster control
(B) Stable condition	(F) Disaster eruption	(J) Disaster termination
(C) Early warning signs	(G) Confrontation begin	(K) Reconciliation
(D) Problem occurs	(H) Confrontation develops	(L) Stable condition

Figure 7.1 Stages in a disaster

- *Problem development*: the early warning signs were missed or ignored. No action was taken, so the problem has developed into a crisis leading to a disaster.

- *Disaster eruption*: a disaster has erupted and disaster management efforts have begun.

- *Confrontation begins*: the disaster management effort doesn't give quick results, because the problem was left to develop into a disaster, so the first task of the disaster management team is to stop the fast escalation of the disaster.

- *Confrontation develops*: the second task for the disaster management team is to contain the disaster within certain boundaries to limit the range of losses.

- *Disaster control*: the disaster is now under control and the disaster management team are trying to reduce the area of disaster to arrive at the core of the disaster.

- *Disaster termination*: the core of the disaster has been reached and handled successfully.

- *Reconciliation*: a quasi-stable condition is reached; the cause of the disaster has been identified and cleared, and normal service is now being established.

- *Stable condition*: the stable condition is reached; the disaster and its effects are over. The loss and damage are assessed; lessons are learnt and recorded.

7.2 FACTORS AFFECTING DISASTER

7.2.1 The effects of the system

How the system affects the disaster depends on the type of disaster. The most prominent causes of the disaster are not technical problems but fires, natural disasters (lightning, floods, earthquakes, tornadoes, etc.) and software. Table 7.2 illustrates the different types of disaster, their causes, probability of occurrence, symptoms, and their effects on the system, the company buildings, and the earth surface around the company. It also illustrates the primary and secondary action needed to handle the disaster situation.

7.2.2 The effects of people

The organization's most valuable resource is its people. The disaster has an impact on the place, the processes and the people. So the reaction of

Table 7.2 Disaster analysis: causes, symptoms, actions

Disaster type	Causes	Can it be predicted?	Symptoms	Effect on the system	Effect on the buildings	Effect on earth surface	Primary immediate action	Secondary action
Technical	Technical malfunction	Mostly yes	Improper operating parameters	Partial or total interruption of service	Generally none	Generally none	Activate alternative service Begin disaster confrontation	Restore main service Begin disaster assessment
Earthquake	Underground activity	Mostly no	Ground shakes, fails or ruptures	Partial or total interruption of service Utility lines are broken	Damage or collapse of buildings	Damage or collapse of dams, roads and bridges	Activate alternative service Search, rescue, medical assistance Initial disaster assessment	Restore main service Do final disaster assessment Repair and reconstruct buildings
Cyclone	Weather High-speed wind	Generally yes	High-speed winds	Partial or total interruption of service Utility lines are broken	Damage to buildings and roads	Damage to buildings, roads and power lines	Activate alternative service Search, rescue, medical assistance Initial disaster assessment	Restore main service Do final disaster assessment Repair and reconstruct buildings

(continued overleaf)

Table 7.2 (*Continued*)

Disaster type	Causes	Can it be predicted?	Symptoms	Effect on the system	Effect on the buildings	Effect on earth surface	Primary immediate action	Secondary action
Flood	Servere rain	Generally yes	Rivers flooding	Partial or total interruption of service Utility lines are broken	Undercuts foundations Buildings are covered by silt	Erodes top soil, deposits silt and changes the course of streams	Activate alternative service Search, rescue, medical assistance Initial disaster assessment	Restore main service Do final disaster assessment Repair and reconstruct buildings
Volcano	Underground activity	Mostly yes	Ground shakes Blast and lava flows	Partial or total interruption of service Destruction of everything it hits	Buildings may be damaged, destroyed or set on fire	Inundates all in its path Causes fire Damages machinery	Evacuation Activate alternative service Search, rescue, medical assistance Initial disaster assessment	Restore main service Do final disaster assessment Repair and reconstruct buildings

Lightning	Static discharge from charged clouds	Generally yes	Thunderstorms and lightning flashes	Partial or total interruption of service Can cause a break in power lines	Buildings and equipment may be damaged, destroyed or set on fire	Localized	Activate alternative service Search, rescue, medical assistance Initial disaster assessment	Restore main service Do final disaster assessment Repair and reconstruct affected areas
Fire	Improper human activity Chemical reaction	Generally yes	Fire erupts Smoke develops Fire spreads	Partial or total interruption of service Destruction of everything it hits	Buildings and equipment may be damaged or destroyed	Localized	Activate alternative service Begin firefighthing Search, rescue, medical assistance	Restore main service Begin disaster assessment Repair and reconstruct affected areas

people to a disaster is very important, since the disaster management plan depends on them for its success. It is therefore important to maximize the people's resistance to any impacts, consequences, and effects resulting from the dangerous events. This is done through motivation, training and disaster management drills.

The staff should be trained how to respond to fire alarms, bomb threats and other emergencies. They should vacate the building once a fire alarm sounds, they should not return to pick up any items they might have left, and they should report the incident to the responsible supervisor.

The staff must be well trained to discharge their responsibilities within the disaster communications plan; they must know who to report to and where to find the emergency communications equipment.

Unfortunately, people are also a major cause of disaster. To reduce the probability that people cause a disaster, we have to resolve conflicts and confrontations within the organization's employees and we have to increase internal and external security measures within the organization.

7.2.3 Priorities when handling a disaster

Here are the management's priorities when handling a disaster:

- Protect the staff and other personnel
- Maintain customer services and traffic flow
- Protect premises, facilities, equipment, programs and supplies
- Protect vital documents

To be able to handle these priorities properly and efficiently, certain security procedures should be applied. These procedures can be divided into three categories: necessary, required and desired.

Necessary security measures

Necessary security measures include procedures for fire control, alarm handling, evacuation, handling natural disasters, handling technical disasters, and alarm reporting. The necessary security measures must be implemented as soon as possible according to a comprehensive disaster management plan approved by the management.

Required security measures

Required security measures include procedures and reasonable precautions to prevent serious disruption of the operation of the organization as a whole and in all individual areas, e.g. engineering operations (outside plant, switching, transmission), engineering planning (short-term and long-term), financial operations (sales, marketing) and employee relations (motivation, conflict resolution). Required security measures must be implemented in order of priority and as soon as possible.

Desirable security measures

Desirable security measure include procedures and reasonable precautions to prevent serious disruption in any area of the organization and to keep the business under smooth control. Desirable security measures should be implemented as circumstances allow; but if a hazard is detected in any area, the confrontation level is dealt with as a reasonable measure or maybe as a required measure, depending on the seriousness of the hazard.

7.3 REQUIREMENTS FOR HANDLING DISASTER

7.3.1 Objectives of a disaster recovery plan

- To identify vulnerabilities in the system

- To protect human life

- To minimize the effect of a disaster on the business

- To provide procedures for use during a disaster

- To develop protection policy against internal and external disaster threats

7.3.2 Elements of a disaster recovery plan

1. Identify and determine the vulnerabilities and potential risks facing the organization.

2. Determine or estimate the priorities of the identified vulnerabilities and potential risks.

3. Determine the occurrence probability of each disastrous event identified.

4. Identify ways and means to remedy the high-priority vulnerabilities and disastrous events.

5. Determine the expected cost of the disaster on the organization.

6. Conduct an inventory of available resources.

7. Determine the expected investment needed to remedy the effect of the disaster.

8. Obtain management's approval for the plan.

9. Recruit and train the disaster management team on their tasks according to the plan.

7.3.3   Examples of disaster recovery

Recovering a computer business

The range of disaster recovery services might include the initiation of a fully equipped portable computer/communications system at the alternative site containing the following facilities:

- Telephone and personal computer or video terminal at each workstation.

- Network operating server-based disk, Windows 2000, Windows NT, Windows 3.x, MS-DOS and Macintosh-based PCs; personal and high-speed laser printers, dot matrix line printers; T-1 connection and modems for dial-in/dial-out and network server/remote access.

- Fax machines and PC-based fax servers with scanners.

- Industry-standard tape formats.

- LAN and asynchronous connections.

- Telephone PBX with forwarding to workstations and DID access.

- Uninterruptible power systems (UPS) for key network and workstation equipment.

The scope of computer crime

In March 2000 the Computer Security Institute (CSI) in San Francisco published its fifth annual computer crime and security survey. Here are some of the findings:

- 71% of respondents detected unauthorized access by insiders.

- 59% cited their Internet connection as a frequent point of attack.

- 38% cited their internal systems as the frequent point of attack.

As in previous years, the most serious financial losses occurred through theft of proprietary information (66 respondents reported US$ 66 708 000) and financial fraud (53 respondents reported US$ 55 996 000).

Survey results illustrate that computer crime threats to large corporations and government agencies come from both inside and outside their electronic perimeters, following the trend in previous years. These findings confirm that the threat from computer crime and other information security breaches continues unabated and that the financial toll is mounting.

Types of computer crime

CSI's survey respondents detected a wide range of attacks and abuses. Here are some examples:

- 25% detected system penetration from the outside.

- 27% detected denial-of-service attacks.

- 79% detected employee abuse of Internet access, e.g.
  — downloading pornography
  — downloading pirated software
  — inappropriate e-mails

- 85% detected computer viruses.

The survey asked some questions about electronic commerce over the Internet. Here are some of the results:

- 19% suffered unauthorized access or misuse within the last twelve months.

- 32% didn't know if there had been unauthorized access or misuse.

- 35% of those acknowledging attack reported from two to five incidents.

- 19% reported ten or more incidents.

- 64% of those acknowledging an attack reported website vandalism.

- 60% reported denial of service.

- 8% reported theft of transaction information.

- 3% reported financial fraud.

7.3.4 Time to clear the disaster

A critical parameter is the period of time to clear a fault behind a disaster. The longer the disaster continues, the higher the costs of the remedy. The cost need not always be financial, sometimes it may be political. Table 7.3 shows the rules for handling faults in a high-power AM radio broadcasting station (composed of two parallel units). It indicates the kind of tension the shift engineer suffers when he or she is faultfinding a disastrous event, knowing that there may be severe penalties. When handling these situations, special preparations should be made to insure the faultfinding progresses quickly towards its objective. Failure to do so would lead to more disastrous events, as explained in Section 11.3.

7.3.5 Criticality considerations

A modern telecommunications organization cannot survive very long without continuous access to its business information and data. To provide business continuity in the event of a disaster, the organization's prime planning and recovery goal must be to insure availability of a working alternative immediately after the disaster has occurred. One of the most

Table 7.3 The penalties of clearing a time-critical disaster

Will fault interrupt the service?	Fault duration (min)	Fault handler	Postclearance documentation	Postclearance action
No	<1	Technician	Report to shift engineer	If report is okay then no further action
No	<5	Shift engineer	Report to chief engineer	If report is okay then no further action
No	>5	Chief engineer	Report to director general	If report is okay then no further action
Yes	<1	Shift engineer	Report to chief engineer	If report is okay then no further action
Yes	<5	Shift engineer Chief engineer	Report to director general	If report is okay then no further action
Yes	>5	Shift engineer Chief engineer	Report to director general and board chairperson	The shift engineer is investigated by the legal authorities for negligence

important elements is to insure that telephone and fax traffic is redirected automatically from a disaster-struck facility to the chosen alternative facility. Table 7.4 can be used to evaluate the criticality of a disaster.

How to use the criticality table

- Complete panels A and B for each type of disaster that you have identified from the vulnerability search.
- Add the criticality ratings of the two tables and complete panel C.
- The resulting criticality index relative to the needed resources will indicate the sort of action taken:

  1 = immediate action for implementation

  2 = implement as soon as possible

  3 = implement after a review of the data has been affirmative

  4 = implement if extra funds are available

  5 = review the disaster management plan to ascertain the criticality rating

Table 7.4 Criticality of a disaster

A. *DURATION/IMPACT*				
Facility			Location	
Type of disaster				
Evaluator's name			Position	
Disaster impact (financial loss)	Duration of the disaster			
	Very long	Long	Medium	Short	Instantaneous
Very high	10	9	8	7	6
High	9	8	7	6	5
Medium	8	7	6	5	4
Low	7	6	5	4	3
Very low	6	5	4	3	2

B. *PRIORITY/IMPACT*				
Facility			Location	
Type of disaster				
Evaluator's name			Position	
Disaster impact (financial loss)	Occurrence probability of the disaster			
	Very high	High	Medium	Low	Very low
Very high	10	9	8	7	6
High	9	8	7	6	5
Medium	8	7	6	5	4
Low	7	6	5	4	3
Very low	6	5	4	3	2

C. *RESOURCES/IMPACT*				
Facility			Location	
Type of disaster				
Evaluator's name			Position	
Criticality index	Disaster recovery resources needed			
	Very high	High	Medium	Low	Very low
Very high >17	1	1	1	1	1
High (17–15)	2	2	2	2	2
Medium (14–12)	3	3	3	3	3
Low (11–8)	4	4	4	4	4
Very low <8	5	5	5	5	5

7.3.6   Equipment and staff to clear the crisis

Management resources inventory

Several areas should be investigated and inventory lists prepared so the disaster management team obtain a full picture about the resources already available at the organization:

- In a telecommunications organization these might include the telephone systems units (line plant, switching and transmission, and power equipment).

- In electronic data processing (EDP) equipment this would include peripherals, modems, multiplexors and network hardware.

- In the TV and broadcasting facility this would include TV and sound broadcasting studios, transmitters (TV, AM, FM, SW), microwave program links, antennas, outside TV and outside broadcast vans.

- The networks utilized (LAN, WAN, MAN, etc.) and the cable types used in the networks.

- The premises would include the building layout, with the electrical wiring diagrams, security systems, air-conditioning, fire detection and firefighting equipment, all clearly marked.

- The external facilities would include cable routes, circuit designator numbers, assignments of cable pairs, conduit routes, vault locations, and the jack, block and pin assignments at the main distribution frame (MDF).

- All the above categories need well-trained staff (engineers and technicians), test and repair equipment, service manuals, and other documentation to make a success of the recovery plan.

7.4   DISASTER MANAGEMENT RESPONSIBILITIES

7.4.1   Reviewing the disaster recovery plan

Table 7.5 could be useful in reviewing the plan, since it details all the elements needed to relate the disaster to the level of impact, and the proposed action. If the action proposed is not compatible with the impact, then the plan needs to be modified.

7.4.2   Testing the disaster recovery plan

The reason for testing the plan is to examine its validity by trying out methods for use in a real disaster. The disasters provided for can be fire, theft, lightning, malfunction in one piece of equipment, partial or even complete destruction of computing facilities, traffic congestion on main routes, and many more.

Table 7.5 Review of the disaster recovery plan

Disaster location or effect	Cause	Area hit	Level of impact			Proposed action	Is it okay?	Remarks
			Severe	Moderate	Mild			
Damage to terminal equipment								
Localized damage								
Damage to a single facility								

(continued overleaf)

Table 7.5 *(continued)*

Disaster location or effect	Cause	Area hit	Level of impact			Proposed action	Is it okay?	Remarks
			Severe	Moderate	Mild			
Major damage to the building								
Damage to the whole organization								
Other damage (specify)								

Table 7.6 Testing the disaster recovery plan

Disaster	Area hit	Measure taken	Test results			Suggested remedy	Remarks
			Okay	Pass	Fail		
Technical							
Fire							
Lightning							
Arson							
Fraud							
Employee error							
Earthquake							
Cyclone							
Flood							
Volcano							
Other (specify)							

The objective of a disaster plan is to insure effective protection against any event that may cause partial or complete loss of service in an organization. The test is carried out by simulating a disaster, in a specific area, to test how the disaster management team will handle it.

The most effective test method is the limited disaster test, in which only one part of the system is subjected to a simulated disaster; although this confines the scope of the test, it does indicate the appropriateness of the disaster prevention measures.

Table 7.6 is a useful way to document the results of each test, and to suggest remedies to any problems that arise during the simulation process.

7.4.3 Modifying the disaster management plan

The success criteria of the simulated test are that it achieves effective disaster management and service restoration. The objectives here are:

- To gather information on the performance of the disaster management team.

- To determine the worth of the simulation project itself.

The test parameters and sequence can be varied to achieve the anticipated results. New tests can be proposed that get closer to the set objectives.

7.5 COST OF A TELECOMMUNICATIONS DISASTER

7.5.1 Cost of disaster recovery and prevention

Business-related costs

Business-related costs could be thought of as the drop in revenue due to the interruption of service, yet far more important are their effects on the public image of the organization.

If the service is interrupted for reasons not readily perceived by the customers, the long-term effects might be devastating and far more costly than any new and updated disaster management plan. If the service is based on a contract, the customer will be tempted to cancel it — in case of repeated interruptions — and this will increase the losses of the organization due to the inadequacy of its disaster preparedness.

The effect of repeated disasters reflects negatively on the cash flow of the organization, since it will be forced to pay extra funds to clear the repeated disasters that it did not cater for. The organization's share value might drop and a financial disaster might ensue. Table 7.7 illustrates the costs due to business-related issues.

Table 7.7 Business-related costs

Disaster		Duration (hours)	Service interruption			Financial losses (lost revenue)			Remarks
			Total	Partial	Minimal	High	Medium	Low	
1	Technical fault	<0.5							
	Technical fault	<1							
	Technical fault	>1							
2	Arson	<0.5							
	Arson	<1							
	Arson	>1							

(continued overleaf)

Table 7.7 (*Continued*)

	Disaster	Duration (hours)	Service interruption			Financial losses (lost revenue)			Remarks
			Total	Partial	Minimal	High	Medium	Low	
3	Fraud	<0.5							
	Fraud	<1							
	Fraud	>1							
4	Human error	<0.5							
	Human error	<1							
	Human error	>1							

Table 7.7 (Continued)

	Disaster	Duration (hours)	Service interruption			Financial losses (lost revenue)			Remarks
			Total	Partial	Minimal	High	Medium	Low	
5	Fire	<0.5							
	Fire	<1							
	Fire	>1							
6	Thunderstorm	<0.5							
	Thunderstorm	<1							
	Thunderstorm	>1							

(continued overleaf)

Table 7.7 (Continued)

	Disaster	Duration (hours)	Service interruption			Financial losses (lost revenue)			Remarks
			Total	Partial	Minimal	High	Medium	Low	
7	Tornado	<1							
	Tornado	<5							
	Tornado	<24							
8	Hurricane	<1							
	Hurricane	<5							
	Hurricane	<24							

Table 7.7 (Continued)

	Disaster	Duration (hours)	Service interruption			Financial losses (lost revenue)			Remarks
			Total	Partial	Minimal	High	Medium	Low	
9	Flood	<1							
	Flood	<5							
	Flood	<24							
10	Earthquake	7–8[a]							
	Earthquake	5–6[a]							
	Earthquake	4–5[a]							

[a]Earthquake strength (Richter scale) not duration in hours.

The results of CSI's fifth annual computer crime and security survey, announced in March 2000, indicate that 90% of respondents detected attacks and 74% acknowledged financial losses due to computer breaches. Some 42% were willing and/or able to quantify their financial losses. The losses from these 273 respondents totaled US$ 265 589 940, while the average annual total over the last three years was US$ 120 240 180.

Financial losses in eight of twelve categories were larger than in any previous year. Furthermore, financial losses in four categories were higher than the combined total of the three previous years. For example, 61 respondents quantified losses due to sabotage of data or networks for a total of US$ 27 148 000. The total financial losses due to sabotage for the previous years totaled only US$ 10 848 850. As in previous years, the most serious financial losses occurred through theft of proprietary information (66 respondents reported US$ 66 708 000) and financial fraud (53 respondents reported US$ 55 996 000).

Service recovery costs

Service recovery costs include the actual repair, costs of equipment or buildings, spare parts consumed or purchased due to the disaster, purchase of new equipment or renting of facilities to replace the damaged items, costs due to personnel casualties, and all other related costs. Table 7.8 can be used to calculate these costs for selected disasters.

Legal costs

The responsibility of the disaster management team is to insure the disaster management plan protects the employees, the organization's resources and any vital documents. Unsatisfied customers may file lawsuits demanding compensation, and the management will become responsible when it is discovered that the plan doesn't adequately meet these requirements.

The managers who have not taken proper action to minimize the effect of a possible disaster are exposing themselves to legal action, if and when a disaster happens and causes casualties and losses that could reasonably have been prevented (Chapter 10). Table 7.9 can be used to calculate these costs for selected disasters.

Profile of financial disaster

To arrive at the total cost of a disaster, we have to add up the business-related costs, the service recovery costs and the legal costs. Using Table 7.10 the disaster management team can grade the disaster according to its impact, then they can decide the financial priorities and determine the proper action in each criticality category.

Table 7.8 Service recovery costs

	Disaster	Duration (hours)	Service interruption			Recovery costs (repair + damage + new/rented equipment + other)			Remarks
			Total	Partial	Minimal	High	Medium	Low	
1	Technical fault	<0.5							
	Technical fault	<1							
	Technical fault	>1							
2	Arson	<0.5							
	Arson	<1							
	Arson	>1							

(continued overleaf)

Table 7.8 (continued)

	Disaster	Duration (hours)	Service interruption			Recovery costs (repair + damage + new/rented equipment + other)			Remarks
			Total	Partial	Minimal	High	Medium	Low	
3	Fraud	<0.5							
	Fraud	<1							
	Fraud	>1							
4	Human error	<0.5							
	Human error	<1							
	Human error	>1							

Table 7.8 (continued)

	Disaster	Duration (hours)	Service interruption			Recovery costs (repair + damage + new / rented equipment + other)			Remarks
			Total	Partial	Minimal	High	Medium	Low	
5	Fire	<0.5							
	Fire	<1							
	Fire	>1							
6	Thunderstorm	<0.5							
	Thunderstorm	<1							
	Thunderstorm	>1							

(continued overleaf)

Table 7.8 (continued)

	Disaster	Duration (hours)	Service interruption			Recovery costs (repair + damage + new/rented equipment + other)			Remarks
			Total	Partial	Minimal	High	Medium	Low	
7	Tornado	<1							
	Tornado	<5							
	Tornado	<24							
8	Hurricane	<1							
	Hurricane	<5							
	Hurricane	<24							

Table 7.8 (*continued*)

	Disaster	Duration (hours)	Service interruption			Recovery costs (repair + damage + new/rented equipment + other)			Remarks
			Total	Partial	Minimal	High	Medium	Low	
9	Flood	<1							
	Flood	<5							
	Flood	<24							
10	Earthquake	7–8[a]							
	Earthquake	5–6[a]							
	Earthquake	4–5[a]							

[a]Earthquake strength (Richter scale) not duration in hours.

Table 7.9 Legal costs

	Disaster	Duration (hours)	Service interruption			Legal costs (fees + compensation)			Remarks
			Total	Partial	Minimal	High	Medium	Low	
1	Technical fault	<0.5							
	Technical fault	<1							
	Technical fault	>1							
2	Arson	<0.5							
	Arson	<1							
	Arson	>1							

Table 7.9 (continued)

	Disaster	Duration (hours)	Service interruption			Legal costs (fees + compensation)			Remarks
			Total	Partial	Minimal	High	Medium	Low	
3	Fraud	<0.5							
	Fraud	<1							
	Fraud	>1							
4	Human error	<0.5							
	Human error	<1							
	Human error	>1							

(continued overleaf)

Table 7.9 (continued)

	Disaster	Duration (hours)	Service interruption			Legal costs (fees + compensation)			Remarks
			Total	Partial	Minimal	High	Medium	Low	
5	Fire	<0.5							
	Fire	<1							
	Fire	>1							
6	Thunderstorm	<0.5							
	Thunderstorm	<1							
	Thunderstorm	>1							

Table 7.9 (continued)

	Disaster	Duration (hours)	Service interruption			Legal costs (fees + compensation)			Remarks
			Total	Partial	Minimal	High	Medium	Low	
7	Tornado	<1							
	Tornado	<5							
	Tornado	<24							
8	Hurricane	<1							
	Hurricane	<5							
	Hurricane	<24							

(continued overleaf)

Table 7.9 *(continued)*

	Disaster	Duration (hours)	Service interruption			Legal costs (fees + compensation)			Remarks
			Total	Partial	Minimal	High	Medium	Low	
9	Flood	<1							
	Flood	<5							
	Flood	<24							
10	Earthquake	7–8[a]							
	Earthquake	5–6[a]							
	Earthquake	4–5[a]							

[a]Earthquake strength (Richter scale) not duration in hours.

Table 7.10 Profile of financial disaster

	Disaster	Critically	Business loss (1)	Financial loss (2)	Legal loss (3)	Total loss $(1+2+3)$	Action needed
1		Very serious					Immediate
2		Serious					Immediate
3		Moderate					Urgent
4		Mild					Needed
5		Minor					Deferred

7.5.2 Criticality rating and recommended action

The criticality rating prioritizes the recommended action when a disaster erupts. There are five categories for criticality:

- *A situation that requires very serious attention*: here the situation is very critical to the well-being of the whole organization. In this case disaster recovery must be a major priority to the organization. Because loss due to interruption of service will be potentially very serious, immediate action is required to restore the service as soon as possible.

- *A situation that requires serious attention*: here the situation is critical to the well-being of the organization. A disaster that may affect part of the organization must also be handled immediately, because the loss of business will affect the financial health of the whole organization.

- *A situation that requires moderate attention*: here the organization is susceptible to significant threat that may lead to disaster. The occurrence probability of this threat can either lower the criticality rating if it is highly improbable, or raise it if it is highly probable. In the median case the threat is supposed to inflict minor harm to the organization; urgent action is recommended to prevent an increase in the occurrence probability.

- *A situation that requires mild attention*: here the organization can probably withstand the disaster, because the disaster is mild or because the organization has very good disaster management.

- *A situation that requires minor attention*: here the threat does not seem to pose any danger to the organization, so there is no need for any action to be taken.

7.5.3 Insurance cost

Insurance coverage is needed to safeguard the organization when a disaster occurs. It can be divided into:

- Insurance coverage for equipment and facilities owned by the organization.

- Insurance coverage for equipment and facilities rented or leased by the organization.

- Insurance coverage for records and documents owned by the organization.

- Insurance coverage for emergency equipment the company might need in case of a disaster.

- Insurance coverage for third-party liability.

- Insurance coverage for loss of business revenue due to a disaster (loss-of-profit insurance).

Owned equipment

Table 7.11 illustrates the cost of insurance on the equipment owned by the organization. It contains all the necessary data for the disaster management team to calculate the effect of a disaster that can be handled by insuring the equipment at risk.

Rented or leased equipment

Table 7.12 illustrates the cost of insurance on the equipment rented or leased by the organization. It contains all the necessary data for the disaster management team to calculate the effect of a disaster that can be handled by insuring the equipment which will be rented when a disaster occurs.

Loss of business

Table 7.13 illustrates the cost of insurance against loss of business due to service interruption at the organization's facilities. It contains all the necessary data for the disaster management team to calculate the effect of a disaster that can be handled by insuring the organization, its managers and service staff against loss of business due to service interruption when

Table 7.11   Cost of insurance on owned equipment

	Insurance company		Address				
	Telephone		Contact person				
	Items insured	Original price ($1000)	Date installed	Replacement value	Insurance value	Annual premium	Policy number
1							
2							
3							
4							
5							
6							
7							

a disaster occurs. For comparison it also indicates the cost of the alternative solution.

Third-party liabilities

Table 7.14 illustrates the cost of insurance against third-party liabilities due to service interruption at the organization's facilities. It contains all the necessary data for the disaster management team to calculate the effect of a disaster that can be handled by insuring the organization, its managers and service staff against third-party liabilities due to service interruption when a disaster occurs. For comparison it also indicates the cost of the alternative solution.

Table 7.12 Cost of insurance on rented or leased equipment

Leasing company			Address				
Telephone			Contact person				
Insurance company			Address				
Telephone			Contact person				
	Items leased	Lease terms	Date leased	Annual cost of lease	Lessor liability	Annual premium	Policy number
1							
2							
3							
4							
5							
6							
7							

Table 7.13 Cost of insurance against loss of business

Insurance company	Address						
Telephone			Contact person				
	Details of business lost	Estimated loss ($1000)	Cost of alternative solution	Insurance value	Policy date	Policy number	Annual premium
1							
2							
3							
4							
5							
6							
7							

Table 7.14 Cost of insurance against third-party liabilities

Insurance company		Address					
Telephone		Contact person					
	Details of liabilities	Estimated loss ($1000)	Cost of alternative solution	Insurance value	Policy date	Policy number	Annual premium
1							
2							
3							
4							
5							
6							
7							

Total insurance cost

Insurance company A		Address
Telephone	Contact person	
Insurance company B		Address
Telephone	Contact person	
Insurance company C		Address
Telephone	Contact person	
Insurance company D		Address
Telephone	Contact person	

Comp-any	Details of insured items and liabilities	Esti-mated loss ($1000)	Annual insurance policy premium				Total annual premium
			Owned equip-ment	Rented equip-ment	Loss of business	Third-party liabilities	
A							

(*continued overleaf*)

(*continued*)

Company	Details of insured items and liabilities	Estimated loss ($1000)	Annual insurance policy premium				Total annual premium
			Owned equipment	Rented equipment	Loss of business	Third-party liabilities	
B							
C							
D							
All	Totals						

Table 7.15 Cost-effectiveness of recovery plan for different disasters

Diaster area	Failed equipment	Replacement cost ($1000)		Loss due to failure at present			Loss with upgraded redundant system			Financial comparison	Remarks
		Same system	Redundant system	Time to recover	Extra cost	Business loss	Time to recover	Extra cost	Business loss		
Damage to terminal equipment											
Localized damage											
Damage to a single facility											

(continued overleaf)

Table 7.15 (continued)

Diaster area	Failed equipment	Replacement cost ($1000)		Loss due to failure at present			Loss with upgraded redundant system			Financial comparison	Remarks
		Same system	Redundant system	Time to recover	Extra cost	Business loss	Time to recover	Extra cost	Business loss		
Major damage to the building											
Damage to the whole organization											
Others (specify)											

7.5.4 Cost-effectiveness analysis

A disaster management and recovery plan is developed to minimize the disaster losses suffered by an organization, losses that might affect employees, equipment, buildings and customers. The success of the plan depends on recognizing the relative importance of the organization's resources and assets. Not all resources are equally important or even equally susceptible to disaster. As we have seen, there are priorities for the occurrence of a disaster in certain areas of the organization.

The elements of a disaster recovery plan should be selected with complete awareness of the interrelationships between critical (susceptible) system functions and their related resources. The cost-effective protection of a telecommunications system must depend on the following ideas:

- Identifying the relative importance of each element in the telecommunications system.

- Determining or properly estimating the occurrence probability of an undesired event that can develop into a disaster.

- Fully studying the consequences and possible side effects of the identified disasters, to prevent a domino effect (chain reaction) or disaster proliferation.

- Measures to prevent or minimize the occurrence probability of a disaster.

- Measures to reduce or minimize the losses encountered from a disaster.

Table 7.15 may help to identify the cost-effectiveness of recovery plans for different types of disaster.

7.6 SECURITY AND POLITICAL ISSUES

People play a large part in the creation of disasters. They may commit crimes, vent grievances or express political sympathies. The disaster management plan should address these security issues carefully. Here are things to consider:

- Security of the premises:
  — building security
  — site security
  — fire prevention
  — firefighting
  — disaster preparaton
- Security of the employees:
  — carelessness

- — smoking
- — inexperience
- — harmful intent
- — hackers

- Security of the equipment:
  - — power surges
  - — water damage
  - — fire damage
  - — cable cuts

- Security of the network:
  - — program errors
  - — modem access

- Security of the software:
  - — viruses
  - — time bombs
  - — vendor trapdoors
  - — poor firewalls

Political broadcasters and similar organizations need a hot standby that's permanently available, especially in developing countries.

SUMMARY

Disaster recovery is primarily about detecting symptoms. Detecting the early warning signs of a disaster means observing any slight change in the normal operating parameters of the system. If these early warning signs are ignored, a disaster may be imminent.

Have a disaster management plan with clear objectives and all the necessary elements. Observe the criticality considerations and the maximum clearance time.

The cost of a disaster may be broken into business costs, service recovery costs and legal costs. The cost of insurance is divided into insurance on owned and leased equipment, insurance against loss of business, and insurance against third-party liabilities. Remember the security and political issues.

REVIEW QUESTIONS

1. Define the following terms: disaster prevention plan, disaster management plan, disaster recovery plan.

2. Give some examples from your work experience of how you identified the basic reasons for some disasters.

3. Explain some of the factors that can make telecommunications disasters happen more frequently. Say how you rectified them at your workplace.

4. What are the requirements for handling a telecommunications disaster? Describe some disasters at your workplace.

5. Explain how to calculate the cost of a disaster. Estimate the cost of a disaster that happened at your workplace.

6. Explain how to estimate the total cost of insurance against loss of business. Give your explanation by referring to a disaster at your workplace.

7. Investigate security and political issues at your workplace.

8

Telecommunications Systems and their Vulnerability

OBJECTIVES

- Answer questions at the end of the chapter.
- Discuss what it means to do an upgrade.
- State the outcomes of technical vulnerability searches at your workplace.
- Outline some financial vulnerabilities at your workplace.
- Discuss how to forecast financial failure and comment on your own workplace.
- Show how to assess administrative vulnerability at your workplace.

8.1 GENERAL CONSIDERATIONS

System vulnerability search, analysis and rectification aims to insure the system does not fail and cause service interruption accompanied by financial and other losses. The main items are discussed in Chapter 6.

8.1.1 Upgrading facilities and buildings

Facilities and building design has a major role to play in the disaster development, since a properly designed building will help prevent the quick development of fire, a major cause of disaster. Here are the possible measures; they are discussed in Chapter 9:

- Purpose grouping of buildings
- Space separation between buildings

- Ignition prevention
- Compartmentation
- Space route design
- Isolation
- Segregation
- Constructional component integrity
- Evacuation of occupants

Table 8.1 might be useful in deciding how to upgrade the facilities and buildings, since it takes the possible vulnerability and the resulting disaster with its associated loss, and relates it to the cost and effectiveness of the upgrade and the proposed system. When the technical problems are transformed into financial problems, the picture is a little bit clearer and a decision can be safely arrived at.

8.1.2 Upgrading equipment reliability

Equipment reliability can be upgraded before purchase by selecting the higher-reliability option; this might be very expensive (Figure 5.1). Or it can be upgraded after purchase by using a redundant system

Table 8.1 Decision to upgrade facilities and buildings

Vulner ability	Resulting disaster			Proposed system upgrading			Decision
	Priority	Impact	Cost (loss)	System	Cost	Effect- iveness	

(a)
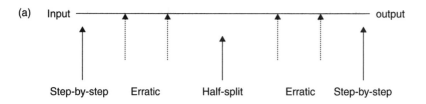

(b)
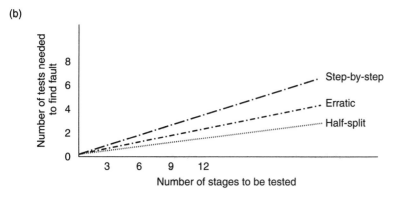

Figure 8.1 Fault detection: (a) where to begin, (b) how many tests

(Figures 5.3, 5.4 and 5.5). Efficient maintenance keeps all equipment healthy (Section 8.2.2).

8.1.3   Upgrading the network configuration

Why develop the network?

- To reduce the cost of interruption.
- To give the network better disaster management performance.
- To prepare the network for the new technology options.
- To improve network survivability and restoration.

Rules for developing a network

- Keep reinstallation costs for new circuits as low as possible.
- Use new technology that can adapt to the existing technology.
- Use the available resources as much as you can.
- Avoid duplicating the circuits, but create redundant routes to increase the network's disaster-handling capability.
- Use any shared resources, since this will help reduce the cost, but ascertain their reliability.

- Use easily replaceable technology, because of the fast pace of technological advances in the telecommunications field.

A regional telecommunications office

During disaster a regional telecommunications office must have several options to route the traffic:

- Microwave or digital radio communications facilities.

- Fiber-optic routes.

- Satellite communications circuits.

- Very small aperture terminals (VSATs).

- Cellular telephony facilities.

- Dual power supply supplied from two different utilities (if possible).

- Standby diesel generator.

- Transportable mainframe in case the fire disrupts the mainframe.

- Transportable telephone exchange.

It should be possible to take the transportable telephone exchange and connect it to the customer lines through the access hole, in case the whole regional office is destroyed. The transportable exchange cannot cater for all the telephone lines in the regional office but will provide service to urgently needed facilities.

The choice between the options will depend on their availability, reliability, ease of use, initial and running costs, and whether there are any legal or operational constraints. They will not be cost-effective for a small PBX serving a remote rural area in a developing country, since the cost of interruption will be far less than the cost of disaster recovery.

8.1.4 Security of telecommunications networks

The security of telecommunications networks covers four main concepts:

- *Integrity*: to insure the system can be assessed only by authorized users or entities.

- *Availability*: to insure the system responds efficiently.

- *Control*: to insure the network managers are able to control user access to certain data.

- *Maintenance*: to maintain the physical connections, the communications equipment and the logical processes defined by the software and procedures.

A secure radio telecommunications system must prevent intruders from reading or modifying the data being transmitted or stored. The security falls into three main categories:

- *Organizational*: such as security rules and management priorities.

- *Technical*: such as the technical means and mechanisms to insure network security.

- *Physical*: such as buildings and the communications methodology.

Security measures

There are four main security function groups that are used for radio telecommunications:

- *Authentication*: to insure the user who attempts to be connected is genuine.

- *Confidentiality*: to insure the information relating to the user will not be listened to or heard by a third party.

- *Data integrity*: to insure the exchanged information will not be modified during transmission.

- *Nonrepudiation*: to insure a given user did actually transmit or receive the message.

The relaxation of security arrangements allows the success of attacks and other security problems. For mobile networks, the use of a radio interface, if decoupled from the fixed physical links of the telecommunications network, leads to security and confidentiality problems.

Security breaches

Technical fraud

- *Rechipping*: reprogramming the electronic chip that contains a terminal's identity number.

- *Cloning*: devising a fraudulent terminal by using the identity of a genuine terminal that is already in use. The fraudulent terminal can make calls without being charged, since the bill will go to the genuine terminal.

Economic fraud

- *Fraudulent subscription*: subscribing under a false name and address; or the operator loses track of the subscriber, who continues to use their subscription rights.

- *Terminal theft*: theft of calls if the terminal is not protected by a password.

Spying and sabotage

- *Passive attack*: the information exchanged between the subscribers is intercepted by eavesdropping.

- *Active attack*: the intent is to alter or destroy information held on the network management database.

Security services and mechanisms

The confidentiality of radio telecommunications systems can be assured by adopting the following measures, according to the required level of security:

- *Level 1*: ciphering is provided on the radio interface and authentication mechanisms are added to the identification process.

- *Level 2*: this uses end-to-end ciphering, not just air interface ciphering; it secures the confidentiality of professional transactions. End-to-end ciphering is applied between telecommunications entities, not just between the system and the mobile.

- *Level 3*: this meets the confidentiality constraints of the military or strategic user. It employs special reinforced ciphering and authentication methods.

8.1.5   Preparing for positive action

To prepare your employees for positive action, they must be motivated to behave in a positive manner (Chapter 5). One of the best ways to motivate people is to delegate some work to them. Delegation can be a risky business, but the risks only occur if it is not handled properly. Most managers are not happy to delegate, because:

- They fear losing control.
- They regret giving up jobs they enjoy doing.
- They believe they can cope with the job better by themselves.

Yet if the manager knows the proper delegation procedure, they will see to it that the job is done well. Here are the appropriate steps:

1. Define the responsibilities:
   - Define responsibilities for the delegated person.
   - Establish key result areas for them to achieve.
   - Review their progress regularly.

2. Set standards of work performance:
   - Set standards in relation to:
     — quality
     — quantity

— cost
— time
- Make the standards measurable.
- Devise a procedure to resolve shortfalls.

3. Set clearly defined targets to be achieved:
   - Make the targets challenging.
   - Praise people who perform well.

4. After delegating, do these six things:
   - Inform other employees about the change.
   - Ask other employees to cooperate with the delegated person.
   - Set a series of meetings to monitor the progress of the task.
   - Agree the method of monitoring with the delegated person.
   - Be available to give advice.
   - Actively check up on key points.

5. After delegating, try not to:
   - Interfere and double-check the delegated person's work.
   - Override the delegated person's decisions.
   - Change things behind the delegated person's back.

8.2 ASSESSING TECHNICAL VULNERABILITY

8.2.1 Vulnerability search, analysis and rectification

The essence of technical vulnerability search, analysis and rectification is to insure the effect and impact of any technically oriented disaster will be as low as practically possible.

Technical vulnerability search carefully and honestly identifies all possible vulnerabilities in the telecommunications system. Technical vulnerability analysis determines the levels of disaster in each vulnerability area. Then it considers how to reduce them. Technical vulnerability rectification takes action to reduce the risk of a technically oriented disaster until its level is acceptable.

There are several types of vulnerability depending on the risk they pose to the telecommunications system. Table 8.2 takes the four telecommunications disciplines and details the proposed search, analysis and rectification work needed for each type.

8.2.2 Maintenance procedures

A successful maintenance program requires management commitment and a proper attitude towards the maintenance activities and staff. Uninterrupted operation and good disaster handling depend on a successful maintenance program, so commitment and attitude are very important.

Table 8.2   Technical vulnerability

Type	Description	Expected disaster	Rectification method
		A. LINE PLANT	
Poses extreme risk (i.e. water flooding access holes)	Found at once (check the access holes)	Easy to analyze	Work should start immediately to rectify
Unnecessary risk	Easily found	Easy to analyze	Work should start immediately to rectify
Poses apparent risk	Needs proper organized search work	Needs careful analysis	Work should start immediately to rectify
Hidden risk	Needs careful search work for concrete data	Needs careful analysis of data	Work should start immediately to rectify
Poses minor and remote risk	Needs proper search for its probability of occurrence	Needs careful analysis of the possible impact and the loss involved	Work should start to consider the appropriate action
Poses no risk to the organization	Search to gather data	Analyze to ascertain that it is not a vulnerability	No action taken if it is not a vulnerability
		B. SWITCHING	
Poses extreme risk (air-conditioning inoperative)	Found at once (check the system)	Easy to analyze	Work should start immediately to rectify
Poses unne-cessary risk (interfacing problems)	Easily found (check interface specs)	Easy to analyze	Work should start immediately to rectify
Poses apparent risk (surpassing the exchange's capacity to carry traffic)	Needs proper organized search work	Needs careful analysis	Work should start immediately to rectify

Table 8.2   (*continued*)

Type	Description	Expected disaster	Rectification method
	B. SWITCHING (*continued*)		
Hidden (no maintenance is performed on exchange)	Needs careful search work for concrete data	Needs careful analysis of data	Work should start immediately to rectify
Poses minor and remote risk	Needs proper search for its probability of occurrence	Needs careful analysis of the possible impact and the loss involved	Work should start to consider the appropriate action
Poses no risk to the organization	Search to gather data	Analyze to ascertain that it is not a vulnerability	No action taken if it is not a vulnerability
	C. TRANSMISSION		
Extreme risk (no alternative route)	Found at once (check the traffic per route)	Easy to analyze	Work should start immediately to rectify
Poses unnecessary risk (alternative route is unreliable)	Easily found (check reliability figures and compare)	Easy to analyze	Work should start immediately to rectify
Poses apparent risk (traffic variations erratic)	Needs proper organized search work	Needs careful analysis of the profile	Work should start immediately to rectify
Hidden (routing software problems)	Needs careful search work for concrete data	Needs careful analysis of data	Work should start immediately to rectify
Poses minor and remote risk	Needs proper search for its probability of occurrence	Needs careful analysis of the possible impact and the loss involved	Work should start to consider the appropriate action

(*continued overleaf*)

Table 8.2 *(continued)*

Type	Description	Expected disaster	Rectification method
		C. TRANSMISSION (continued)	
Poses no risk to the organization	Search to gather data	Analyze to ascertain that it is not a vulnerability	No action taken if it is not a vulnerability
		D. ENERGY	
Poses extreme risk (no alternative energy supply)	Found at once	Easy to analyze	Work should start immediately to rectify
Poses unne-cessary risk (unreliable main energy supply)	Easily found	Easy to analyze	Work should start immediately to rectify
Poses apparent risk (reliability of alternative supply is low)	Needs proper organized search work	Needs careful analysis	Work should start immediately to rectify
Hidden (main and alterna-tive supply routes are the same)	Needs careful search work for concrete data	Needs careful analysis of data collected	Work should start immediately to rectify
Poses minor and remote risk	Needs proper search for its probability of occurrence	Needs careful analysis of the possible impact and the loss involved	Work should start to consider the approppriate action
Poses no risk to the organization	Search to gather data	Analyze to ascertain that it is not a vulnerability	No action taken if it is not a vulnerability

Maintenance principles

Operation and maintenance are virtually inseparable; we evaluate our activities and learn from our experience, we identify shortcomings and prepare a solution, we evaluate the solution and put it into practice, then we start all over again.

Unplanned maintenance

Unplanned maintenance, or nonsystematic maintenance, is carried out according to no predetermined plan. The equipment continues to run until it fails, then it is either repaired or replaced. There are two types of unplanned maintenance:

- *Corrective maintenance* is carried out after a failure has occurred; it aims to restore the failed item to its normal functioning.
- *Emergency maintenance* has to be performed immediately to avoid serious consequences.

Planned maintenance

Planned maintenance, or systematic maintenance, covers four basic systems:

- *Preventive maintenance* is carried out at predetermined intervals or corresponding to prescribed criteria; it is intended to reduce the probability of failure or the performance degradation of an item. Preventive maintenance has to be done cost-effectively.
- *Scheduled maintenance* attempts to forestall a breakdown by identifying, on a historical basis, the duration of the failure interval exhibited by a component or system; it is carried out at predetermined time intervals.
- *Condition-based maintenance* is preventive maintenance initiated to deal with a condition discovered by routine or continuous monitoring.
- *Reliability-centered maintenance* is a systematic approach to preventive maintenance; it uses quantitative assessments to help schedule maintenance tasks and plans.

Effects of improper maintenance

- Reduced benefits
- Higher costs
- Reduced profitability

Factors influencing maintenance

- *Location*: the location of the organization must make it easy to accommodate the technicians and procure any spare parts.
- *Equipment*: look at the condition of the equipment and facilities; the older they are, the higher the probability of interruption and the higher the cost of maintenance.
- *Personnel*: telecommunications equipment is rapidly updated with ever more sophisticated items, and this increases the need for highly

trained technicians. Developing countries find it particularly hard to recruit people if they buy state-of-the-art systems.

- *Misuse and abuse*: equipment is usually misused through ignorance; abuse may be down to discontent, vandalism and many other factors.
- *Funds*: developing countries usually buy local services in local currency and imported spares in foreign currency; they often have difficulty in securing foreign currency. This is an important point.

A model maintenance system

Maintenance program

- What is to be done
- How it is to be done
- How often it is to be done
- Who is responsible

Personnel

- Explain to each team member what is expected from them
- Prepare a well-written job description
- Give proper training on maintenance activities
- Give proper on-the-job training
- Make maintenance training a continuous activity

Maintenance management

- Delegate responsibility
- Perform follow-up activities
- Train staff effectively and efficiently
- Define maintenance program details
- Recruit proper supervisors
- Insure all items are available before maintenance begins
- Establish maintenance records

Parts

- Purchase parts on time and store them properly
- Decide when, where and how to buy items, and how much you will pay
- Establish good warehouse buildings and records

Records

Records are very important for making annual budgets of the needed spares; everything must be well documented. The minimum record for each item must contain the following data:

- Name of equipment or facility
- Serial number, type and class
- Date of installation
- Cost of purchase
- Name and address of manufacturer
- Details of any service organization
- How to obtain service manuals
- Physical location of the unit
- Previous operating problems
- Previous overhauls
- Quantity and cost of spare parts used

Establishing a maintenance plan and schedule

Critical units and service windows

Study your plant (telecommunications center, radio or TV transmitting station, information technology center, etc.). Determine its nature and classify it into units, then construct a process flow diagram. Study the consequences of failure in each unit and estimate the cost of lost service. Study the operation plan:

- Determine the critical plant units.
- Rank the critical units according to their failure costs.
- Identify the service window where a failure will cause the lowest cost.
- Determine how the existing maintenance plan can cope with your findings.

Constituent items

Classify the plant into critical units and noncritical units. To save time and effort, only the critical units will be completely classified.

Effective maintenance procedure

For each unit determine the most effective maintenance procedure from the cost and safety viewpoints. This will be tabulated to show certain parameters such as:

- The item to be maintained
- When it will need maintenance
- The timing and period of maintenance

- Whether to take it out of service
- The periodicity of maintenance
- The specific maintenance action needed

Work plan

The plan will depend on the flowchart of the plant, whether it is a series flow or batch. The series flow is the most difficult since we have to consider the whole plant as one unit when scheduling our maintenance activities. The availability of redundant units simplifies the job.

Maintenance schedules

Online maintenance work can be scheduled independently for each unit or item, since it will not affect the normal operation of the plant. Offline work can be done during service windows, or as part of an agreed shutdown. Always aim to carry out the maintenance work smoothly, and with the minimum of service loss.

Corrective maintenance guidelines

Even with proper routine or preventive maintenance, a system may still fail through some unexpected event, due to random failures which are very difficult to monitor. Use reliability analysis to predict random failures then plan your spares and staff accordingly.

Maintenance for an earth station

The size of the earth station will affect the maintenance philosophy. Staff at a small earth station will have a wider range of responsibilities than staff at a large earth station, where a higher degree of specialization is required. The maintenance responsibilities at a medium-size earth station can be divided into six areas:

- The antenna system
- RF wideband equipment
- IF equipment
- Terminal equipment
  — baseband
  — multiplexor
  — frequency convertor
- Power equipment
- Control and supervisory, and test facilities

Each member of the maintenance team should be specialized in one of those areas and have a good idea about at least two of the other areas. When an item is found faulty, make a quick decision from these three options:

- Repair it on site
- Send it to the central repair site
- Send it to the manufacturer

Prepare a predictive maintenance program that includes these five factors:

- The reliability of each component in the earth station, and the time it has been working properly.
- The availability of faultfinding and repair staff with the required expertise.
- Type, cost and number of spares needed according to the reliability predictions made.
- The necessary test equipment needed to perform the tests.
- The technical documentation about the earth station, incorporating all the amendments.

8.2.3   Faultfinding procedures

Locating faults in systems

Three methods are used to locate faults in electronic systems (Figure 8.1):

- *The erratic method*: measurements are made according to the experience of the technician with other faults they have encountered on this equipment or similar equipment.
- *The step-by-step method*: measurements are made in a systematic way; there are two variations:
  — working from input to output
  — working from output to input
- *The half-split method*: measurements are made in a systematic way, beginning from the middle stage; this saves time in detecting the failure.

System faultfinding aids

There are many faultfinding aids; choose the ones that suit your type of equipment. For digital systems the most important aid is the diagnostic program to determine the faulty subunit or subassembly. The fault is usually in a printed board, and the components are most often surface mounted, so the only way to repair the fault is by replacing the printed board.

Analog and old-fashioned equipment can be diagnosed using many different tools:

- Manufacturer's O&M manual
- Detailed circuit diagram
- Multimeter
- Multifunction generator
- Wideband oscilloscope
- Signal analyzer
- Soldering and desoldering irons
- Aerosol contact cleaner
- Freezer sprays
  — for temperature-dependent faults
- Components and spare parts

Troubleshooting electronic systems

General considerations

The main idea in troubleshooting electronic systems is to restore the system to normal (or quasi-normal) as quickly as possible. So the time factor is very important. Be sure of these vital points:

- Know the details of the faulty circuit well, so you do not spend valuable time in trying to figure out what has happened and why.
- Perform FMEA for this unit so you know what has happened and what components are causing the fault.
- Have the spare parts at hand without any administrative complications.
- Have a backup or emergency circuit at hand, in case the fault will take a long time to clear.

RF transmitters

Even before a fault occurs, make sure you carry out these steps:

1. Perform the necessary maintenance.
2. Perform the necessary tests after the maintenance is complete to ascertain that the transmitter is functioning at peak performance:
   - Tests for AM transmitters:
     — Rated power output
     — Peak power output
     — Power output leveling
        (for multichannel transmitters)
     — DC power input

- — Modulation percentage
- — Modulation distortion
- — Audio frequency response
- — Hum and noise modulation
- — Modulation sensitivity
- — Detection of incidental FM
- — Automatic modulation control (AMC)
- — Carrier shift detection and measurements
- — Frequency
- — Parasitic oscillation
- — Improper neutralization
- Tests for FM transmitters:
  - — Power output measurement
  - — Modulation/deviation
  - — Peak deviation
  - — Percent deviation dissymmetry
  - — Modulation sensitivity
  - — Voice modulation
  - — Modulation density
  - — Modulation bandwidth
  - — Audio frequency response
    - in-band
    - out-of-band
  - — FM hum and noise level
  - — Audio distortion
  - — Frequency

3. Take (and record) measurements at all available test points when the transmitter is in optimum condition.

4. If there is a fault, use the faultfinding techniques described earlier. Measure at the prescribed test points and compare the results with the values you obtained when the transmitter was okay.

5. Detect the faulty component or subsystem and repair or replace it as appropriate.

6. Check for proper performance then return the item to service.

Coaxial cable system

Capacitance testers can detect a break in the coaxial cable; the method is quick and simple.

1. Measure the capacitance in picofarads (pF) at one end of the cable (point 1) and call it C_1.

2. Measure the capacitance in picofarads at the other end of the cable (point 2) and call it C_2.

3. The distance of the break from point 1 is given by

$$\text{Distance} = \frac{C_1}{C_1 + C_2} \times \text{cable length}$$

Radio relay links

Carry out these steps before faultfinding begins:

1. Perform the necessary maintenance.

2. Perform the necessary tests after the maintenance is complete to ascertain that the link is functioning at peak performance:
   - Measure the system's gain in decibels (dB). This equals the transmitter power output (dBm) minus the minimum receiver input power (dBm) for the required quality objective.
   - Calculate the free path loss.
   - Calculate the fade margin; it depends on the multipath effect, the terrain sensitivity, the required reliability and a constant.
   - Measure the signal-to-noise ratio (SNR) of the receiver.
   - Measure the carrier-to-noise ratio (CNR).
   - Find the noise figure (NF):

$$\text{NF (dB)} = 10 \log \left(\frac{\text{input SNR}}{\text{output SNR}} \right)$$

3. Take (and record) measurements at all available test points when the transmitter is in optimum condition.

4. If there is a fault, use the faultfinding techniques described earlier. Measure at the prescribed test points and compare the results with the values you obtained when the transmitter was okay.

5. Detect the faulty component or subsystem and repair or replace it as needed.

6. Check for proper performance then return the item to service.

The manufacturer normally supplies visual indicators (in a special rack) which help the operation and maintenance staff in their faultfinding activities. Manufacturer's flowcharts also help to simplify the faultfinding process; here is an example.

The manufacturer has supplied the visual indicator rack and flowcharts, and an engineer has found that indicator 10 in the receiver section is lit. The engineer will refer to the flowchart (Figure 8.2) to identify the faulty unit. The faulty item is identified and isolated. The faulty unit is then replaced with a spare and the faulty unit is either repaired at the organization's workshop or returned to the manufacturer for repair.

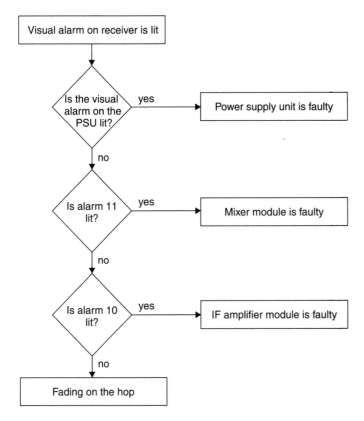

Figure 8.2   Link receiver: fault detection flowchart

Digital telephone systems

Digital telephone systems are discussed on page 300.

Intelsat satellite system

The main causes of interruption in an earth station can be divided into six categories:

- *Failure in the satellite*: the Intelsat organization will transfer the traffic to the spare satellite.

- *Failure in the earth station*: by using redundancy the fault will not affect the operation of the earth station; the alternative equipment will be switched into the circuit immediately, perhaps automatically, unless there is damage to the antenna itself.

- *Natural phenomena*: these include passage of the sun or moon into the antenna beam of the downlink earth station. Natural phenomena will be known in advance and proper contingency plans prepared to reduce their effects.

- *Interference*: the source of the interference must be identified and eliminated.

- *Natural disasters*: a hurricane strike may cause loss of pointing accuracy in the antenna or serious damage to the earth station building. An earthquake may be even more destructive. Take account of natural disasters during the design phase of the earth station, and increase its disaster-handling capabilities.

- *Human error*: human error is the most serious category; it lies behind bad maintenance, poor faultfinding expertise, negligence, mistakes due to poor training, or poor ergonomic design.

Fiber-optic systems

Fiber-optic systems are usually supplied with several faultfinding aids to help the technician:

- A diagnostic program with alarms that indicate a fault in a specific plug-in unit.

- Visual indicators (lamps) to indicate the state of each subunit.

- Alarm surveillance channels giving access to all fault information for all parts of the system.

- A man–machine interface to assist in quick isolation of the fault.

- Machine–machine interfaces to send the fault information to the central maintenance monitoring unit.

However, the automatic diagnostic program may not detect some faults. For example, when an alarm indicates there is a fault at one site, the fault at that site can be in the receiver (it has failed), in the transmitter (the output has been reduced so that it was not detected by the diagnostic), or in the outside plant (where a fiber could have been cut). To detect which one of the three possibilities is the cause, some measurements must be made, such as:

- *Measuring the input optical power at the receiver*: if it is okay, then the receiver has failed; if not, then the transmitter is faulty or the outside plant is faulty.

- *Measuring the transmitter output power*: if it is below the specified value, then the transmitter is faulty; if not, then the problem should be with the outside plant fiber.

Before restoring traffic from the protection channel to the repaired channel, it is recommended to verify the performance of the repaired channel. This can be done by testing that the bit error rate (BER) is within specifications (about 10^{-10} per repeater station).

Troubleshooting electrical components

High-voltage cables

Cables can fail for several reasons. Table 8.3 gives the symptoms of the fault and the probable cause in each case.

Medium-voltage XLPE cables

All cables experience hazards during the time between shipment and installation. Insulation made from cross-linked polyethylene (XLPE) has a high risk of damage if the cable is dragged. Yet medium-voltage cables cannot be tested by the high-voltage DC test, so they require a new non-intrusive, non-life-reducing test. When a cable becomes faulty, the first step is to determine the approximate location of the fault; this can be done by one of these methods:

- *Time domain reflectometer*: this detects open circuits short circuits and fault insulation resistance.

- *Bridge method*: this detects resistive faults

- *Impulse current method*: this detects high resistance and flashover faults, plus most of the faults detected by the other two methods

The second step is to determine the exact location of the fault by using a surge generator:

- *Low-voltage surge generator*: apply a few surges in single-shot mode at the lowest possible voltage of the surge generator and listen to the suspected fault area.

- *High-voltage surge generator*: when the first method fails, use a higher voltage from the surge generator and listen to the suspected fault area.

Table 8.3 Cable failures

Symptom	Measurement	Expected cause
Cable inoperative	Impedance	Open-circuit conductor
Low voltage	Impedance	Low resistance ($<10Z_0$)
Very low voltage	Impedance	High resistance ($>10Z_0$)
Flashing	Impedance	Breakdown due to high voltage impulse
Intermittent	Impedance	Occasional breakdown

Table 8.4   Maintenance and repair test voltage specifications

Test waveform	Duration	Test voltage (kV)			
		6.6	11	22	33
VLF	15 min	8	13	25	38
50 Hz	15 min	7	11	22	33
Surge	5 surges	7	11	22	33

When the fault has been detected and the cable repaired, several tests are performed before the cable is inserted back into service:

- *Overvoltage test*: the cable is tested according to the schedule in Table 8.4.

- *Leakage test*: this measures the leakage resistance of all the conductors to earth with a DC voltage of more than 250 volts

When these tests indicate that the cable has been properly repaired, it can be put back into service.

8.3   ASSESSING FINANCIAL VULNERABILITY

Financial reporting is supposed to provide useful information to company managers, creditors and existing potential investors, to help them make rational decisions regarding company operations, investments and credits. The information provided should therefore be comprehensible to those who have some understanding of business and economic activities, and they should study it with reasonable diligence.

Financial reporting mainly focuses on earnings and earnings components, and it aims to give a good indication of the organization's present and continuing ability to generate favorable cash flows. It also provides information about the organization's financial performance during a specified period (yearly or quarterly). It shows how management has discharged its responsibilities towards the owners or shareholders. And it provides information about the financial resources of the organization, the claims against them, and the effects of the transactions, events and circumstances that change these resources and claims.

Management performance is evaluated by the financial performance of the organization, and vulnerabilities are also detected by analyzing the financial performance of the organization. Investors and creditors can estimate the earning power of the organization, predict future earnings and assess risk by studying the different financial statements.

8.3.1   Basic financial terms

Assets

Assets are probable future economic benefits obtained or controlled by a particular entity, as a result of past transactions or events. Assets can be either current or noncurrent:

- *Current assets* may be in cash or something that will be realized in cash. They are listed on the balance sheet in order of their ability to be converted into cash (liquidity):
  - *Cash*: cash in hand and negotiable checks; cash is the most liquid of all assets.
  - *Marketable securities*: debt instruments of the government and other companies that can be readily converted into cash.
  - *Accounts receivable*: monies due on customer accounts, arising from sales or services rendered.
  - *Inventories*: the balance of goods on hand; in a manufacturing company they may include raw materials, work in progress, finished goods and supplies.
  - *Prepaids*: expenditures made in advance for goods or services.

- *Noncurrent or long-term assets* are assets that do not qualify as current assets; they are divided into:
  - *Tangible assets*: such as land, buildings, machinery.
  - *Investments*: such as stocks and bonds of other companies.
  - *Intangibles*: nonphysical assets such as legal rights.

Liabilities

Liabilities are probable future sacrifices or economic benefits arising from present obligations of a particular entity to transfer assets or provide services to other entities in the future, as a result of past transactions or events. Liabilities are divided into current liabilities and long-term liabilities. Current liabilities include:

- *Payables*: short-term obligations created by the acquisition of goods or services, such as accounts payable, wages payable and taxes payable.

- *Unearned income*: monies collected before the services have been performed.

Long-term liabilities are liabilities due in a period exceeding one year; they fall into two categories:

- Liabilities relating to financing agreements, such as notes payable and bonds payable.

- Liabilities relating to operational obligations, such as pension obligations, deferred taxes and service warranties.

Equity

Equity is the residual interest in the assets of an entity that remains after deducting liabilities. Thus, equity = assets − liabilities.

Investments by owners

Investments by owners are increases in equity of the enterprise resulting from transfers to the enterprise from other entities of something of value to increase ownership interests in it.

Distribution to owners

Distribution to owners is a decrease in equity of the enterprise resulting from transferring assets, rendering services or incurring liabilities by the enterprise to the owners.

Comprehensive income

Comprehensive income is the change in equity or net assets of a business enterprise during a period, resulting from transactions and other events from nonowner sources.

Revenues

Revenues are the inflows or other enhancements of assets of an entity resulting from delivering or producing goods, rendering services within the entity's ongoing major operations.

Expenses

Expenses are outflows or other consumptions, the consuming of assets, or incurrence of liabilities of an entity resulting from delivering or producing goods, rendering services within the entity's ongoing major operations.

Gains

Gains are increases in equity or net assets resulting from peripheral or incidental transactions of an entity.

Losses

Losses are decreases in equity or net assets resulting from peripheral or incidental transactions of an entity (except those that result from expenses or distribution to owners).

What period should the accounts cover?

The only accurate way to account for the success or failure of a business is to accumulate all transactions from the day it opens to the day it is wound up. This obviously cannot be done while business is still active. Accounting for the success or failure of the business in midstream will therefore

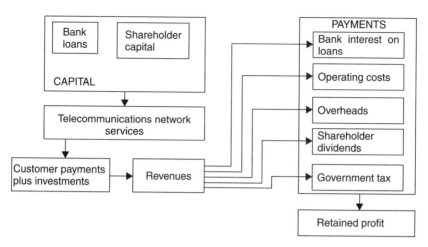

Figure 8.3 Financial accounts: basic components

involve inaccuracies, because many transactions and commitments will be incomplete.

Even when accounting is done for a specific period (yearly or quarterly) the inaccuracies still exist. Suppose an entity carries accounts receivable on the balance sheet, it can only account for them accurately when the receivables are actually collected. It may also have some outstanding obligations at any particular time, and these obligations cannot be accurately accounted for until they are met.

As managers and engineers, we are only interested in the overall performance of the organization. We want to be sure that no disaster will occur due to a grave financial decision. Thus we can live with these inaccuracies and endeavor to study the available financial statements to detect any obvious dangers in the short or long term. Figure 8.3 shows the basic components of the financial accounts.

8.3.2 Basic financial statements

The basic financial statements are the balance sheet, the income statement (profit and loss), and the statement of cash flow.

Balance sheet

The purpose of the balance sheet (Figure 8.4) is to show the financial condition of the organization on a particular date. It consist of assets, which are the resources of the organization, liabilities, which are the debts of the organization, and the stockholder equity, which is the owners' interest in the organization. The assets must always equal the liabilities plus the equity. Here are some relevant points when presenting a balance sheet:

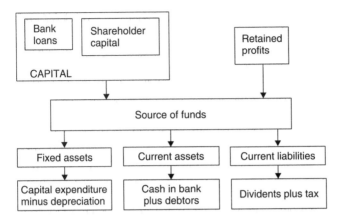

Figure 8.4   Telecommunications balance sheet: basic components

- The valuation of assets at cost: it may be impossible to determine the market value or replacement cost of many assets; do not assume their balance sheet value equals their current valuation.
- Method used for asset valuation: inventories may be valued differently from business to business, or from product to product; this makes them difficult to compare.
- Not all items of value to the organization are included as assets, e.g. outstanding employees.

Income statement (profit and loss)

The income statement (Figure 8.5) is a summary of revenues, expenses, gains and losses, ending with a net income for a particular period of time. It summarizes the results of operations for an accounting period. Net income is closed to the retained earnings account in the stockholder equity section of the balance sheet; this is necessary for the balance sheet to balance.

Statement of cash flow

The statement of cash flow (Figure 8.6) details the sources and uses of cash during a specified period of time; this must be the same period used for the income statement (usually one year). It has three major sections: cash flows from operating activities, cash flows from investing activities and cash flows from financing activities.

Statement of retained earnings

The statement of retained earnings is not a required financial statement, but it usually accompanies the income statement, balance sheet and statement of cash flows. It is a reconciliation of the retained earnings at the end of the prior period, to the retained earnings balance at the end of the current period. Figure 8.7 illustrates the interrelationships between the balance

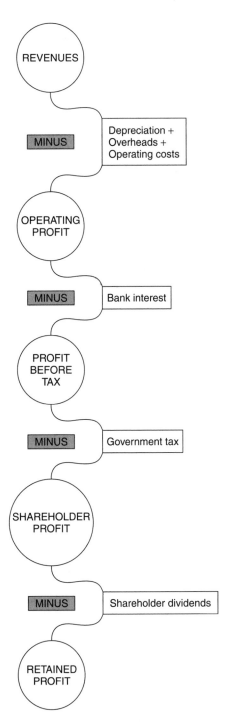

Figure 8.5 Telecommunications income statement: basic components

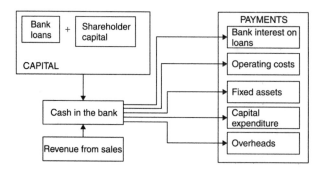

Figure 8.6   Telecommunications cash flow statement: basic components

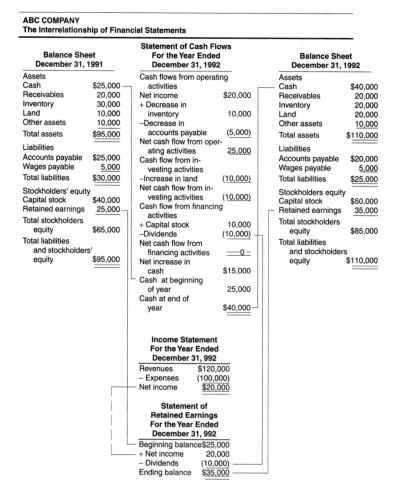

ABC COMPANY
The Interrelationship of Financial Statements

Balance Sheet December 31, 1991		Statement of Cash Flows For the Year Ended December 31, 1992		Balance Sheet December 31, 1992	
Assets		Cash flows from operating		Assets	
Cash	$25,000	activities		Cash	$40,000
Receivables	20,000	Net income	$20,000	Receivables	20,000
Inventory	30,000	+ Decrease in		Inventory	20,000
Land	10,000	inventory	10,000	Land	20,000
Other assets	10,000	–Decrease in		Other assets	10,000
Total assets	$95,000	accounts payable	(5,000)	Total assets	$110,000
		Net cash flow from oper-			
Liabilities		ating activities	25,000	Liabilities	
Accounts payable	$25,000	Cash flow from in-		Accounts payable	$20,000
Wages payable	5,000	vesting activities		Wages payable	5,000
Total liabilities	$30,000	–Increase in land	(10,000)	Total liabilities	$25,000
		Net cash flow from in-			
Stockholders' equity		vesting activities	(10,000)	Stockholders equity	
Capital stock	$40,000	Cash flow from financing		Capital stock	$50,000
Retained earnings	25,000	activities		Retained earnings	35,000
Total stockholders		+ Capital stock	10,000	Total stockholders	
equity	$65,000	–Dividends	(10,000)	equity	$85,000
Total liabilities		Net cash flow from		Total liabilities	
and stockholders'		financing activities	–0–	and stockholders	
equity	$95,000	Net increase in		equity	$110,000
		cash	$15,000		
		Cash at beginning			
		of year	25,000		
		Cash at end of			
		year	$40,000		

Income Statement
For the Year Ended
December 31, 992

Revenues	$120,000
– Expenses	(100,000)
Net income	$20,000

Statement of
Retained Earnings
For the Year Ended
December 31, 992

Beginning balance	$25,000
+ Net income	20,000
– Dividends	(10,000)
Ending balance	$35,000

Figure 8.7   Interrelationships between financial statements (Reprinted, with permission, from C.H. Gibson, 1992, *Financial Statement Analysis*, Southwestern Publishing)

sheet, income statement, statement of retained earnings and statement of cash flows.

8.3.3 Basic analysis of financial data

Various techniques are used in financial analysis; they are designed to emphasize the comparative and relative importance of the data and to evaluate the position of the organization.

Ratio analysis

- *Liquidity ratios* measure the organization's ability to meet its current obligations.

- *Borrowing capacity ratios* measure the degree of protection for suppliers of long-term funds; they indicate the organization's long-term debt-paying ability.

- *Profitability ratios* measure the earning ability of the organization, including the use of assets; when calculating the ratios, we depend only on income that is expected to occur in subsequent periods.

- *Cash flow ratios* indicate liquidity, borrowing capacity or profitability; relatively new, they didn't become required until 1988.

Common size analysis

Common size analysis is a way to express figures in percentages, more meaningful when comparing businesses having different sizes; the figures are brought to a common (percentage) base. There are two types:

- *Vertical* common size analysis takes a figure from one year and compares it with a base selected from the same year.

- *Horizontal* common size analysis takes a figure and expresses it in terms of that same figure in a selected base year.

Comparisons with other data

Figures or ratios are almost meaningless unless they are compared with other data. Trend analysis uses the past history of the business; it looks at a trend in a particular ratio to see whether it is falling, rising or remaining relatively constant. Trend analysis helps to avoid financial disasters. Comparisons between organizations of different sizes may be more revealing than comparisons between organizations of the same size. Look at relative sales, assets or profits. Financial statements may vary from one industry sector to another; see what this reveals. Compile industry averages and industry rankings then see how your business is placed.

Blend your analyses

The information derived from these techniques should be blended to determine the overall financial position of the organization. Financial analysis is a matter of judgment; its primary objective is to identify turning points or major changes in trends, amounts and relationships, and to propose their underlying causes. A negative turning point may be an early warning sign of a significant shift leading to business failure and disaster.

8.3.4 Financial ratios

Liquidity ratios

- Current ratio $= \dfrac{\text{current assets}}{\text{current liabilities}}$

- Cash ratio $= \dfrac{\text{cash equivalent} + \text{marketable securities}}{\text{current liabilities}}$

- Sales to working capital $= \dfrac{\text{sales}}{\text{average working capital}}$

Borrowing capacity ratios

- Debt ratio $= \dfrac{\text{total liabilities}}{\text{total assets}}$

- Debt to equity ratio $= \dfrac{\text{total liabilities}}{\text{shareholder equity}}$

- Debt to tangible net worth $= \dfrac{\text{total liabilities}}{\text{shareholder equity} - \text{intangible assets}}$

Profitability ratios

- Net profit margin (return on sales) $= \dfrac{\text{net income}}{\text{net sales}}$

- Total asset turnover $= \dfrac{\text{net sales}}{\text{average total assets}}$

- Return on assets $= \dfrac{\text{net income}}{\text{average total assets}}$

- Operating income margin $= \dfrac{\text{operating income}}{\text{net sales}}$

- Return on operating assets $= \dfrac{\text{operating income}}{\text{average operating assets}}$

- Sales to fixed assets $= \dfrac{\text{net sales}}{\text{average net fixed assets}}$

- ROI $= \dfrac{\text{net income} + (\text{interest expense})(1 - \text{tax rate})}{\text{average long-term liabilities} + \text{average equity}}$

- Gross profit margin $= \dfrac{\text{gross profits}}{\text{net sales}}$

Notes

- *Net income* means net income before minority share of earnings and nonrecurring items.

- *Net profit margin* measures the net income dollars generated by each dollar of sales.

- *Total asset turnover* measures the activity of the assets and the ability of the organization to generate sales through the use of assets.

- *Return on assets* measures the organization's ability to utilize its assets to create profits.

- *Sales to fixed assets* measures the organization's ability to make productive use of its property, plant and equipment by generating sales dollars.

- *Return on investment (ROI)* measures the relationship between the income earned and the capital invested.

- *Gross profit* represents the cost of the product sold during the period.

Cash flow ratios

- A ratio measuring the organization's ability to meet its current maturities:

$$\frac{\text{operating cash flow}}{\text{current maturities of long-term debt and current notes payable}}$$

- A ratio measuring the organization's ability to cover total debt with the yearly cash flow:

$$\frac{\text{operating cash flow}}{\text{total debt}}$$

- Operating cash flow per share is a ratio measuring the organization's ability to cover cash dividends with yearly operating cash flow; the higher the ratio, the better the ability:

$$\frac{\text{operating cash flow} - \text{preferred dividends}}{\text{cash dividends}}$$

8.3.5 Forecasting financial failure

There are many studies on the use of financial ratios to forecast financial failure. These studies try to isolate individual ratios or combinations of ratios that may forecast failure. Here are two models; one of them uses a single ratio, the other uses five.

The univariate model

William Beaver developed the univariate model in October 1968. It uses just a single variable. The Beaver study indicated that the following ratios were the best for forecasting financial failure:

- Cash flow/total debt
- Return on assets
- Debt ratio

Using these ratios, Beaver computed the mean value of thirteen financial statements for each year before failure; the results indicated that:

- Failed companies have less cash but more accounts receivables.
- When cash and receivables are added together, it obscures the differences between failed companies and successful companies; this is because the differences in their cash and receivables work in opposite directions.
- Failed companies tend to have less inventory.

The multivariate model

Edwin I. Altman developed the multivariate model to predict bankruptcy. His model uses five financial ratios that are weighted to maximize the predictive power of the model. The model produces an overall discriminant score, called a Z value:

$$Z = 0.12X_1 + 0.014X_2 + 0.033X_3 + 0.006X_4 + 0.010X_5$$

where

$$X_1 = \frac{\text{working capital}}{\text{total assets}}$$

$$X_2 = \frac{\text{retained earnings (balance sheet)}}{\text{total assets}}$$

$$X_3 = \frac{\text{earnings before interest and taxes}}{\text{total assets}}$$

$$X_4 = \frac{\text{market value of equity}}{\text{book value of total debt}}$$

$$X_5 = \frac{\text{sales}}{\text{total assets}}$$

X_1 measures the net liquid assets of the company relative to the total capitalization; X_2 measures the cumulative profitability over time; X_3 measures the productivity of the company's assets; X_4 measures how much the company's assets can decline before the liabilities exceed the assets and the company becomes insolvent; X_5 measures the sales-generating ability of the company's assets.

When computing the Z value, the ratios are expressed in absolute percentage terms, e.g. 25% is written 25.0. The lower the Z score, the more likely the company will go bankrupt. In a study conducted over the period 1970–73, a Z score of 2.675 was established as a practical cutoff point. Companies that score below 2.675 are assumed to have characteristics similar to those of past failures.

8.4   ASSESSING ADMINISTRATIVE VULNERABILITY

Administrative vulnerabilities are a major factor in the development of a disaster; they either place the organization under an unnecessarily high risk, or they provide an environment for the disaster to develop, and therefore they unnecessarily endanger the organization. Administrative vulnerabilities stem from the organization's bureaucratic rules and regulations that inhibit the quick action and flexibility needed to deal with dangerous events before they become disasters.

Example 1: Storekeeping rules

A TV and broadcasting organization has demanded that the maximum time allowed to clear a fault (without penalty) must be 5 minutes. The fault must be identified, the defective component or subsystem must be repaired or replaced, and the station must be put back on air — all in less than 5 minutes.

The identification problem is easily solved by studying the station's circuit diagram, simulating faults when the station is off air, and trying to clear them within the time limit. The real problem lies in the storekeeping rules. They require the station engineer to apply to the storekeeper for spares from the stores, and only when the spares are needed (Figure 8.8). The chief engineer must approve this application before the storekeeper will accept it.

After several major delays due to storekeeping, a solution is devised that preserves the rules but allows the shift engineer to request in advance the minimum amount of critical spares to clear all probable faults. The chief engineer approves this application and the spares are stored in a substore whose keys are kept with the shift engineer. So when a fault occurs, the spares are ready and waiting.

Later on there is fault that requires a spare not held in the substore, so the shift engineer must conduct a failure tree analysis to insure the set of critical spares in the substore is complete.

Example 2: Technical manuals

At the same TV and broadcasting organization, six copies of the technical manuals and circuit diagrams are usually delivered by the manufacturer. Five go to the headquarters library and one goes to the chief engineer's library at the transmitting station. Copying the manuals is not permitted,

Figure 8.8   Timing diagram for faultfinding activities

Fault report			
Transmitting center		Transmitting station	
Shift engineer		Shift technician	
Date	Time failed	Time repaired	
Characteristics of the fault			
Breakdown	Intermittent	No RF	No video
No audio	Oscillations	No drive	Degraded performance
Other (specify)			
Cause of the fault			
Equipment failure	Lightning		Maintenance
Other (specify)			
Fault reporting			
Name	Position	Time	Method
Fault clearing			
Name	Position	Time began	
Problem faced			
Repair procedure			
Fault analysis			
Time taken to clear the fault			
Lessons learnt			
Shift engineer's signature		Date	
Chief engineer's recommendations			
Name and signature		Date	
Director general's recommendations			
Name and signature		Date	

Figure 8.9 Fault report for the transmitting station

due to copyright laws, yet the station engineer needs them to prepare for faultfinding and has to ask for the chief engineer's; this creates a major administrative vulnerability. The problem is solved by contacting the manufacturer and obtaining copying approval. As long as they are for sole use of the engineers and technicians, the station may copy circuit diagrams, maintenance routines and faultfinding procedures.

Example 3: Fault reporting

At the same TV and broadcasting organization, the O&M engineers have to submit a technical report to the chief engineer after clearing every fault. There is no unified report format; each engineer adopts their own style. This problem is solved by devising a unified report format (Figure 8.9).

Delays while technical reports receive legal clearance are solved by inviting the legal department to visit the station and meet the O&M engineers.

8.5 SPECIFIC ACTION PLANS

Specific action plans are usually dealt with in relation to the following parameters:

- The response time
- The proposed backup system
- The maintenance regime
- The faultfinding method

The response time should be minimal; there should be very little delay after the disaster is reported. Condition-based monitoring might be a good choice; it will detect problems before they cause disasters.

The following tables might be useful in analyzing the maintenance problems and suggesting solutions. Putting a technical problem in financial terms makes it much easier to reach a decision.

Maintenance problem analysis card

Table 8.5 is a form used for analyzing maintenance problems within the company's departments.

Detailed analysis card

Table 8.6 is a detail enlarged from the maintenance problem analysis card. It analyzes a maintenance problem that seems to have dangerous and disastrous effects on the system. The card details the following information:

- The task or activity where the problem is expected to occur.
- Details of the problems involved.
- The alternative solutions available.
- Evaluation of each solution, giving the estimated cost in each case.
- The recommended solution.
- The expected disaster if the solution is not implemented and an estimate of the losses.

8.5.1 Action plan for telecommunications networks

Preparing the disaster management team

- Train the team members on what they must do before, during and after the disaster.

Table 8.5 Maintenance problem analysis card

Organization name	Facility name		
Date of last analysis	Facility location		
Project code	Prepared by		
Date	Position		
Problem description	Departments concerned		
Projected effect on the system			
Severity	Probability	Cost	Action
Nuisance	Unlikely	Nominal	Defer
Marginal	Probable	Significant	Analyze
Critical	Considerable	Extreme	Investigate
Catastrophic	Imminent	Prohibitive	Immediate

- Choose the proper alternative (hot) site (wireless local loop, PABX, transmission media) for emergency operations in case of a disaster.
- Prepare and distribute a booklet containing the main elements of the disaster management plan plus the location of the alternative outside line plant, PABX or transmission equipment they will use in a disaster.
- Prepare a second alternative site to recover sensitive services if they are disrupted.
- Discuss the benefits of the roll-in PABX; if it is cost-effective, you might use it as the second alternative site.
- Train the team members at the alternative site, regularly and in rotation. Make sure that they become acquainted with the new environment before a disaster occurs and make sure all equipment is in good working order.

Table 8.6 Detailed maintenance problem analysis card

Organization name	Facility name
Date of last analysis	Facility location
Project code	Prepared by
Date	Position
Detailed maintenance problem analysis card number Date	Department concerned
Task/activity	Problem details
Alternative solutions 1 2 3	Evaluation of each solution 1 2 3
Recommendation	
Recommended solution	Estimated cost
Recommended actions 1 2 3	Recommended tasks By whom When How Where
Situation if action is not taken	
Expected disaster	Estimated cost of loss/damage

- Secure communication links to the hot site (wireless local loop, mobile telephones).

- Send team members to disaster-planning workshops where they can share experiences. This will motivate them and have a positive effect on their morale.

Action plan for a digital exchange

Study the exchange block diagram carefully. Figure 8.10 shows a simplified block diagram for the FETEX-150 digital exchange. Now study its maintenance and faultfinding aids:

- The maintenance and operation subsystem (MOS) provides automatic testing and diagnosis.

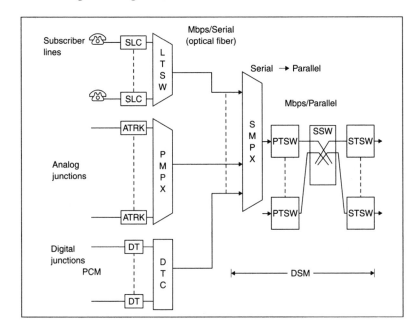

SLC   = subscriber line card
LTSW = line concentrator time switch
ATRK = analog trunk
DT   = dial tone
PMPX = primary multiplexor
DTC   = digital terminal common
SMPX = secondary multiplexor
PTSW = primary line switch
SSW   = space switch
STSW = secondary line switch
DSM   = digital switch module

Figure 8.10 FETEX-150 block diagram (Adapted with kind permission of Fujitsu Limited)

- Autonomous messages and simple commands provide the man–machine interface via a computer screen.

- The technician can operate and maintain the system without too much specialized knowledge.

- A centralized maintenance and operation center (CMOC) can be established to monitor different exchanges.

- Subscriber administration, analysis of historical data, testing and fault processing, all can be done from the CMOC.

Study the software reliability features of the exchange:

- *Hierarchical memory*: memory is organized in three levels (main memory, file memory and magnetic tape memory); the main memory is backed by the file memory, and the file memory is backed by magnetic tape. This makes it very unlikely that data will be lost during a failure.

- *Restart process*: four phases of system restart, each selected according to the fault conditions, minimize the effect of system restart on call processing.

- *Freeze dump*: should a software fault occur, the contents of the temporary data memory are transferred to the file memory, then to the computer screen, and then to magnetic tape. This output data is a useful means of analyzing software faults.

- *Audit*: the software system has a built-in audit facility for early detection of software faults; early detection helps to minimize their effects.

- *Fault detection*: the system employs a rotation method for detecting faults in which doubtful items of equipment are replaced with standbys in a predetermined order. This insures stable and continuous operation.

- *External supervision*: system operation is continuously checked through a periodic artificial call generated by hardware. If the program control becomes abnormal, the artificial call cannot be completed and an emergency circuit activates the system restart.

Study the hardware reliability features of the exchange:

- Speech channels are multiplexed in the digital switching system by time division base, forming highways. Hence equipment failure can influence a large number of calls. Therefore the exchange is provided with full redundancy throughout the system.

- Fault detection, isolation and localization facilities are built into the system in conjunction with the software.

Study the maintenance procedures:

- Maintenance is concentrated in the maintenance and operation subsystem.
- This subsystem provides built-in testing and diagnosis.
- The man–machine interface is through the computer screen.

In case of a technical fault:

- The redundant equipment will be inserted automatically in the circuit.
- Follow the faultfinding procedures suggested by the manufacturer to rectify the fault in the out-of-circuit subsystem.

8.5.2   Action plan for computer/data networks

Preparing the disaster management team

- Train the team members on what they must do before, during and after disaster.
- Choose the proper alternative (hot) site for operations in case of a disaster.
- Prepare and distribute a booklet containing the main elements of the disaster management plan, the transportation arrangements, and the hotels they are going to use.
- Send the team members to the alternative site regularly and in rotation, so they get acquainted with the new environment before a disaster occurs.
- Issue company credit cards to the team members so they can travel immediately to the alternative site in case of a disaster.
- Send team members to disaster-planning workshops.

Four operating problems

Expired contract for the alternative site

When the disaster hits the company and the disaster management team relocates to the alternative (hot) site, it discovers that the service agreement and license for the equipment have expired, because they were not updated after the last test of the disaster management and recovery plan.

Expired software licensing agreements

The team also discovers that the software license agreements at the alternative site have expired. This leaves the team without a working software key.

Team members overwhelmed by logistics

The team members are faced with many logistic problems when moving to the alternative site carrying the tapes. If they were not properly trained on how to act and behave during a disaster, this logistics problem would disrupt the recovery procedure.

Delayed response from the common carriers

Any delay in connecting the alternative site with the necessary telephone lines will render the hot site useless. The disaster management plan must be tested on all the lines, not just a subset. Otherwise, during the disaster, the disaster management team might discover that the common carrier cannot reroute all the required circuits as quickly as expected.

8.5.3 Action plan for TV and broadcasting networks

Preparing the disaster management team

- Train the team members on what they must do before, during and after the disaster.

- Choose the proper alternative (hot) site (TV/sound broadcasting studios and transmitters) for emergency operations in case of a disaster.

- Prepare and distribute a booklet containing the main elements of the disaster management plan, the location of the alternative studio and the transportable transmitter (with easily foldable telescopic antenna) they are going to use to secure the service in case of a disaster.

- Prepare an outside TV and broadcasting van to be located in an alternative site for use as a second alternative site, due to the sensitivity and criticality of the TV and broadcasting media interruption.

- Send the team members to the alternative site, regularly and in rotation, so they get trained in the new environment and they can insure the equipment is in good working order.

- Secure program links to the hot site (microwave link or fiber-optic network) and equip the hot sites with enough recorded material for at least six hours of transmission.

- Send team members to disaster-planning workshops where they can share experiences. This will motivate them and have a positive effect on their morale.

8.5.4 Action plan for electrical power distribution

Preparing the disaster management team

- Train the team members on what they must do before, during and after the disaster.

- Choose the proper alternative (hot) site (substation connected to the electricity grid) for emergency operations in case of a disaster.

- Prepare and distribute a booklet containing the main elements of the disaster management plan and the location of the alternative substation they are going to use to secure the service in case of a disaster.

- Prepare another source of emergency electrical energy for use as a second alternative, due to the sensitivity and criticality of the electricity supply interruption; perhaps it feeds critical business or media.

- Send the team members to the alternative site, regularly and in rotation, so they get trained in the new environment and they can insure the emergency setup will cope with the added load during a disaster.

- Send team members to disaster-planning workshops where they can share experiences. This will motivate them and have a positive effect on their morale.

SUMMARY

Assessing the vulnerability of a telecommunications system begins by identifying the vulnerabilities and upgrading the facilities, buildings, equipment reliability, network configuration and security systems. Important but often overlooked is the motivation of personnel. Carry out vulnerability, search, analysis and rectification then revise maintenance and faultfinding procedures. Engineers and technicians seldom discuss financial vulnerability, yet it plays an important part in disasters. Look at the relevant financial indicators and ratios; understand the forecasting abilities of univariate and multivariate models.

REVIEW QUESTIONS

1. Define the following terms: liability, equity, assets.

2. Discuss the main security concepts for telecommunications networks. How are they applied at your workplace?

3. How can you assess a telecommunications system's technical vulnerability? How would you assess the situation at your workplace?

4. How can you assess a telecommunications system's financial vulnerability? How would you assess the situation at your workplace?

5. Discuss how you could establish a maintenance plan and schedule for the telecommunications system at your workplace. What advantages do you envisage?

6. Explain how to prepare a simple financial statement. How can you detect any financial warning signs by analyzing these statements? What financial ratios would you use?

7. Prepare an action plan to safeguard your organization against any clear disaster that you have previously identified.

9

Safety Considerations

OBJECTIVES

- Answer questions at the end of the chapter.
- Discuss the different aspects of a safe system.
- Outline some safety measures for your workplace that can increase your company's profits.
- Give examples of firefighting, fire control and fire prevention that can be applied at your workplace.
- Describe the nature of lightning strikes and how to eliminate lightning disasters at your workplace.
- Explain how to identify accident-prone personnel and how you can manage them at your workplace.

9.1 THE CASE FOR A SAFE SYSTEM

The safety of telecommunications facilities is very important in disaster management. All telecommunications equipment, including items not exposed to customers, should be designed with safety in mind. We must insure that, under normal operating conditions, all people who install, test, maintain and operate these items will not be subjected to any hazardous situations.

A safety system is the set of procedures, staff and equipment specifically designed to be applied by the organization to increase its safety. The case for a safe system is obvious; many organizations have identified and addressed this need by establishing safety departments, writing safety manuals containing safety procedures and regulations, and by purchasing necessary equipment. Yet the main question is how to evaluate the proposed safety system to insure maximum return on investment, taking into consideration the limited budget available for safety, so that maximum effectiveness is achieved. The important aspects in designing for safety are:

- The Wiring Regulations
- Company system design guidelines
- Testing to meet safety standards

9.1.1 The Wiring Regulations

The main objective of the Wiring Regulations (IEE Wiring Regulations, BS 7671, sixteenth edition, 1992) is to insure the electrical installation is safe. It is intended to protect persons, property and livestock (in locations intended specifically for them) against dangers of electric shock, fire and burns, and against injury from electrically driven mechanical equipment.

The Wiring Regulations are intended for the professional engineer:

- They are a design manual aimed at the professional electrical engineer.
- They do not replace the need for a detailed design specification.
- They are not intended to be a means of training.
- Their scope does not include all possible electrical installations.

The regulations have no direct legal status; they are cited in the Electrical Supply Regulations 1988 and in the Electricity at Work Regulations 1989. They may be cited in a legal contract, but sections of the regulations may *not* be cited in a legal contract.

9.1.2 Design guidelines

Each telecommunications organization has developed and formulated, over the years, a set of design guidelines to be followed in the manufacture, installation, testing, operation and maintenance of telecommunications equipment. These guidelines are concerned mainly with the safety of the equipment, the employees and the customers. They include:

- Protection against hazardous voltages, either internally generated or externally infected as in the case of lightning strikes or contact with power supply lines.
- Protection against mechanical hazards, such as from moving parts, sharp edges or protruding objects.
- Protection against harmful radiation hazards, such as microwave and laser radiation, and ionizing radiation from alpha, beta, gamma and X-rays.

- Protection against excessive heat by using appropriate heat-dissipating arrangements to minimize the temperature rise. Also, for protecting personnel from touching hot spots.

- Protection against the spread of fires, if and when they occur, by using special plastics and cable covers that have flame-retarding properties.

- Protection against the generation of poisonous gases by using cable covers that do not generate poisonous gases when the cable temperature rises beyond normal.

- Any other relevant protection measure.

9.1.3 Testing manufactured equipment

Manufactured telecommunications equipment is tested in special laboratories to insure it adheres to the safety design guidelines. Underwriters Laboratories (UL) was founded in 1894 to test products for electrical and fire hazards. It is an independent self-supporting safety-testing laboratory. A UL listing implies the equipment meets the UL specification, although UL does not accept any liability for the equipment it lists.

9.2 MANAGING SAFETY INCREASES PROFITS

9.2.1 Introduction

In 1979/80 accidental fires cost France about FFr 855 million, which amounts to 0.32% of France's GDP. The loss of production due to road accidents is estimated at FFr 1.4 million per person. If the person is injured only, the cost of lost production equals FFr 100 000. Thus the cost to the economy due to accidents and fire is enormous.

9.2.2 Examples from Zimbabwe

A mining company lost Z$ 4 million (US$ 0.4 million) in 1994 because of an accident that happened due to a cable fire in the mine electrical control board. It was found later that the cable was not properly dimensioned to carry the full load. A colliery lost Z$ 5.5 million (US$ 0.55 million) in damage to assets arising from accidents during the 1995 operational year. It lost about three times that amount in 1992 (known as the killer year). In 1972 the worst disaster happened when 427 miners were buried underground after a shaft collapsed. The colliery experienced a work-related accident once every three days, costing it millions of dollars in damage to assets; that is why the company began to improve the safety measures by introducing the Prevention of Accidents Committee (PAC). The activities of the committee resulted in a drop in the number of fatal accidents.

9.3 SAFETY MANAGEMENT WITHIN AN OVERALL SYSTEM

The essence of management is to organize people, machines and materials to accomplish specific goals. When these goals are properly attained, the managers are adequately rewarded, otherwise they are removed from their positions. The more senior the manager, the greater their responsibilities. Responsibility implies risk; the manager must be willing to accept the risk involved and they should decline to take on risks they are unconfident of handling.

A safety system should be an integral part of the total management system. It consists of three levels:

- *Level 1*: this specifies the alternative system that will be operational if and when the main system fails. The functions and goals of the alternative system are identified and evaluated, available options are discussed and objectives are formulated.

- *Level 2*: the alternative system is divided into subsystems. The objectives, functions, concepts and approaches of each subsystem are discussed and evaluated then objectives are formulated.

- *Level 3*: the alternative subsystem approaches are considered, discussed and evaluated. A detailed design is adopted for each subsystem.

A safety control system is an integral part of the alternative safety system. It has three elements:

- The system's standard should be realistic and achievable; people should be able to monitor its progress.

- The safety control system must allow measurement of its outputs to determine whether the standards and goals are being achieved.

- A feedback system is needed to detect any deviation from the standard and make corrections to insure standards are adequately met.

9.4 FIREFIGHTING, CONTROL AND PREVENTION

Fire is one of the worst catastrophes to affect an organization; the risk of fire and its cost have increased steadily as technology has evolved and become more sophisticated. An exchange building that once held 10 000 lines may now serve over 200 000. Fire damage can take several forms:

- Damage and destruction of material goods may affect items held in store, equipment out in use, essential archives and data, or the organization's premises. The monetary value of the damage can be calculated as primary loss arising from the destruction by fire, and as secondary loss due to the after-effects of the fire. The cost of borrowing must also be included.

- Fire can endanger life and limb, due to burns or panic. People may jump from windows during a fire when they feel they are going to die anyway. Unlike equipment and property, the cost paid by the fire victims cannot be calculated in monetary terms.

- Fire can affect natural and environmental resources; cable fires can produce pollution and cause a hazard to the surrounding population.

9.4.1   The cost of fire

The cost of fire can be divided into losses and expenses. Here are the losses, evaluated as a percentage of the total cost of fire:

1	Cost of direct loss	30%
2	Cost of indirect losses	5%
3	Human losses	5%
	Total	40%

Here are the expenses, evaluated as a percentage of the total cost of fire:

1	Cost of fire prevention	30%
2	Cost of insurance	15%
3	Cost of emergency services	15%
	Total	60%

Notice that the expenses are greater than the losses, yet there is no way to reduce them except by trying to prevent the fire in the first place.

9.4.2   Fire statistics

There are no separate statistics for fire in telecommunications companies, yet we can be guided by the global figure that the cost of fire in a country lies between 0.8% and 0.9% of its gross domestic product.

9.4.3   Fire detection

Introduction

The role of fire detectors is not only to detect fire but to detect it reliably. If someone is smoking a cigarette, a fire detector must not operate due to the heat of their cigarette. On the other hand, if the detector fails to detect a real fire, a disaster is imminent. The sensitivity of the detector must be adjusted appropriately.

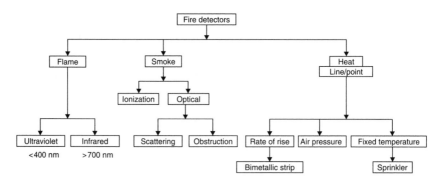

Figure 9.1   Fire detectors

Detector classes

Detectors can be sensitive to heat, smoke and flame (Figure 9.1). Heat detectors have two configurations, point and line. A point detector protects a small area around itself. A line detector consists of several sensitive elements stacked in a continuous line; each part of this line is responsive to heat arising from a developing fire.

Fixed-temperature heat detectors are activated using a bimetallic strip. A fixed-temperature sprinkler detects the fire then extinguishes it. A second type of heat detector is activated when a specified value is exceeded for the rate of rise in temperature of the air and hot gases that flow past.

Smoke detectors respond to smoke from combustion and smoldering fire flames; a sensitive detector must be able to respond to both types of flame. There are two types, ionization and optical. In an ionization detector the air is ionized by a radiation source, some of the resulting ions interact with the incoming smoke particles and the rest are neutralized to produce current. The smaller the current, the higher the count of smoke particles. In an optical detector the smoke interacts with a beam of light by obscuration or scattering:

- *Obscuration detectors* are sensitive to the attenuation of a beam of light shining across a space, caused by the scattering and absorption of the light by smoke particles. A current flows all the time from the photocell detector, but it is reduced by smoke.

- *Scattering detectors* detect the scattered light from suspended smoke particles. Light-emitting diodes are used to increase efficiency and reliability.

Flame detectors recognize radiation from a burning zone. There are two types:

- *Infrared detectors* detect radiation from the flickering flames (in the infrared region) and compare it to a steady radiation. A fire flame is detected when flicker radiation is received.

- *Ultraviolet detectors* detect radiation from the flickering flames (in the ultraviolet region); there is no problem of interference from the sun's ultraviolet radiation since most of it is absorbed by the ozone layer in the earth's upper atmosphere. The ultraviolet radiation is detected by a photocell sensitive to this region of the electromagnetic spectrum.

9.4.4 Choice of a fire detection system

Detector head location

The chosen detector head location will depend on the expected amount of heat energy output from the fire, and on the height of the ceiling. Several parameters should be fixed regarding the siting of detectors and the limit of their applicability, e.g. the maximum distance of any point from the detector, the maximum area that can be covered by one detector, and the distance between each detector and the wall.

Heat detector location

The heat-sensitive element of the detector should be placed away from the ceiling; a distance of 5–10 cm is normal.

Smoke detector location

The detector should be placed at the highest point in the ceiling (in enclosed areas), since this is where all the smoke is trapped (Figure 9.2a). When the ceiling height is greater than 15 m, special precautions should be taken to determine the exact location where smoke will be trapped; this requires a knowledge of the factors which influence smoke production and movement within the building. Smoke control curtains are a good example (Figure 9.2b)

9.4.5 Extinction of fire

Introduction

The rate of burning of any fire can be reduced in three ways:

- Reducing the oxygen reaching the fire
- Reducing the fire flame temperature
- Increasing the activation energy of the fire agent

The most efficient method is to reduce the fire flame temperature.

Extinguishing mechanisms

There are three extinguishing mechanisms: the addition of diluents like water or carbon dioxide; isolation of the fire, perhaps using foam; and

(a)

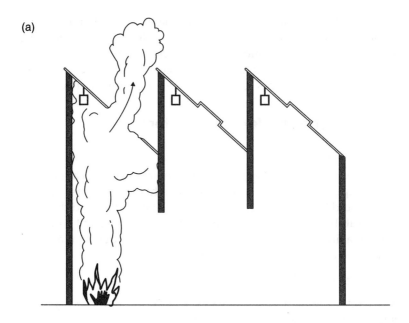

(b)

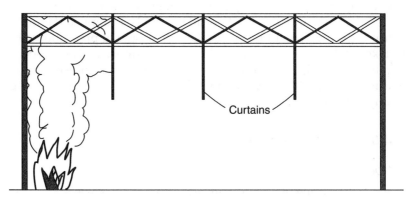

Figure 9.2 (a) Detector location, (b) smoke control curtains

chemical or physical inhibition, perhaps using powder or halons, to break the chain reaction that sustains combustion.

Diluents

- *Water* is sprayed onto the flames; the water evaporates and extracts heat, helping to reduce the flame temperature below its lowest allowable adiabatic value, then combustion stops (Figure 9.3).

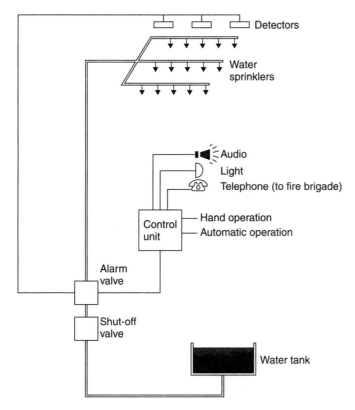

Figure 9.3 Water sprinkler system

- *Carbon dioxide* (CO_2) extinguishes fire by reducing the oxygen content in air so that combustion cannot continue (Figure 9.4)

Isolation

Foam expands over the burning fluid or solid surface, isolating the fire from the oxygen it needs to sustain combustion.

Inhibition

Halons are gases that inhibit combustion by inhibiting its chain reaction. They are very useful in extinguishing fires involving electronic equipment, since they have low toxicity and they are electrically nonconductive. They can extinguish a fire effectively and quickly by total flooding without wetting or leaving a residue (Figure 9.5).

Halons 1301 and 1211 are preferred because they have low toxicity and high extinction effectiveness. They also have a high F factor, which is the ratio of the concentration necessary for extinction to the concentration likely to cause narcosis. When halons are released they can enter areas that

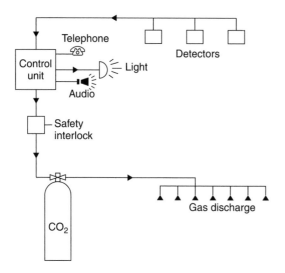

Figure 9.4 Carbon dioxide system

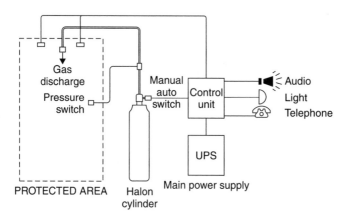

Figure 9.5 Halon gas system

may be inaccessible to water or foam. Total flooding can be implemented in two ways:

- *The central bank method*: the cylinders containing the halon gas are stored outside the premises and a network of fixed pipework (similar to water sprinklers) releases the gas over the whole premises.

- *The modular method*: individual cylinders, at specific locations within the premises, are used to put out the fire. There are four reasons why the modular system is preferred to the central bank method:
  — The modular system is easy to install and it does not occupy large or vital space, since the cylinders can be hung on unused walls.

— The system can be easily enlarged to cover any additional expansion needed.

— The system can be easily relocated to another building or premises.

— The system's operational and maintenance costs are much lower than those of the central bank system.

Alkali metal fires are usually extinguished with dry powder. Dry powder can also act as a diluent to absorb thermal energy from the combustion area. It is very effective in extinguishing liquid-pool fires. The area of the fuel pool determines the amount of powder needed to extinguish it, so the powder extinguishing units range from small hand-held units to large installations of perhaps several tonnes.

Choice of extinguishing system

To be able to extinguish any fire effectively and completely, it is imperative that the extinguishing agent suits the nature of the fire. Table 9.1 gives the suitable agent for a variety of fires.

Fire control systems

Active control systems

Active control systems have four elements:

- *Detection*: see Section 9.4.3.

- *Warning*: the warning depends on the type of building and the occupants; see Section 9.4.6.

- *Firefighting activities*: when fire is detected and the alarm system is activated, several activities can take place successively or simultaneously
  — Begin firefighting by using the portable extinguishing units to try and put the fire out.
  — Activate the automatic firefighting installations.
  — Call the fire brigade; depending on the system used, the call can be made automatically with the activation of the alarm system, or manually by the disaster management team.

Table 9.1 Choice of extinguishing system

Kind of fire	Type of fire extinguishing material				
	Halon	Powder	CO_2	Foam	Water
Paper, wood, textile, fabric	yes	yes	no	yes	yes
Flammable liquids	yes	yes	yes	yes	no
Flammable gases	yes	yes	yes	no	no
Electrical hazards	yes	yes	yes	no	no

- *Direct attack method*: the nature of the attack will depend on the type of building and the occupants. It may differ according to the situation faced. There are many fixed installations that can be used:
  — Water sprinkler
    - spray
    - curtain
    - drencher
  — Carbon dioxide (CO_2) systems
  — Halon gas systems (1301, 1211)
  — Dry powder systems
  — Foam systems
    - ordinary expansion
    - medium expansion
    - high expansion

Passive control systems

Passive control systems rely on the fire endurance characteristics of the building structure and its constructional components (walls, partitions, doors, etc.); they restrict the spread of the fire and leave it to reach an end by itself.

Combination active–passive

A combination active–passive system combines an active system and a passive system.

Smoke control systems

The escape route used by the building occupants in case of fire should be a safe passage at all times; that is why we must insure it will remain usable in an emergency. The escape route must be protected from fire and the resulting smoke. This is important when using passive fire control, since relying on the building's capability to cope with fire will not affect the speed at which smoke develops and moves into the escape route, rendering it unsafe and thus unusable. There are three methods of smoke control: natural ventilation, mechanical ventilation and pressurization.

Natural ventilation

Effective natural ventilation depends on several factors:

- Wind velocity and wind direction at the time the fire erupts; they largely determine the smoke movement's pathways. If these movements are not properly governed, smoke may block the escape routes.

- Stack effects result from the difference between external and internal climate; the difference in temperature causes movements of smoke upwards or downwards until a natural plane is reached. At that point, smoke begins to move laterally.

- The upward force of the rising smoke plume may cause damage to the roof even before fire attacks the roof.

- Mechanical air-movement systems give more effective smoke control than natural ventilation. An effective mechanical ventilation system must be able to:
  — Clear the escape route from smoke
  — Operate efficiently and reliably throughout the firefighting process
  — Operate immediately when a fire warning is activated

Mechanical ventilation

There are two types of mechanical ventilation system (Figure 9.6):

- *Smoke extraction only*: this system uses suction fans to extract smoke. Although the system can be efficient in extracting smoke from the fire area, it might introduce smoke into the escape route, and thus hinder the escape process.

- *Smoke extraction and air inflow*: this system uses suction fans to extract smoke and force air into the affected area. It is a better system than extract only; the critical aspect is choosing where to locate the suction and inflow fans.

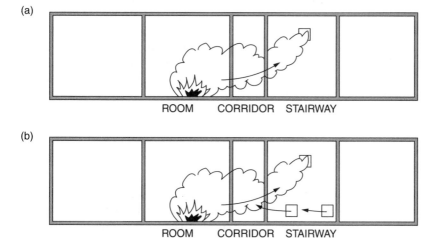

Figure 9.6 Mechanical ventilators; (a) extraction only, (b) smoke extraction and air intake

(a)

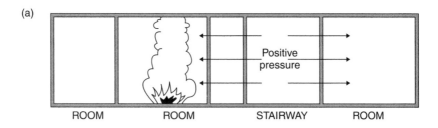

(b)

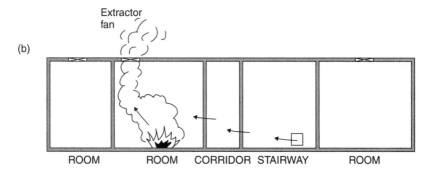

Figure 9.7 Pressurization systems: (a) positive, (b) negative

Pressurization

In this system the escape route is pressurized, fans are used to force air into the escape route to create high pressure there, causing smoke to flow away from the protected escape area. Pressurization can be ineffective if it is not catered for in the original design of the building. When installing a pressurization system in an existing building, make sure it will operate efficiently.

There are two basic types of pressurization system, positive and negative (Figure 9.7). In the positive pressurization system, clear air is pumped into the protected area; this is used to protect the escape route (staircases and corridors). In the negative pressurization system, clear air is sucked from the protected area; this is used to protect the escape route (staircases and corridors).

9.4.6 Alarm systems

Alarm systems are used to alert the occupants of a building that there is a fire danger; they must be designed to suit the building they are supposed to protect and they should be able to communicate a warning signal to the occupants.

Alarms can be aural, visual or a combination of both, since the aural alarm is ineffective for deaf people and the visual alarm is ineffective for blind people. Another important consideration is the function of the building being protected by the alarm; an aural alarm cannot be used in a hospital.

Thus, the choice of an efficient alarm system depends on many factors, such as:

- The people being protected, their type, their perception and their ability to make a quick decision.

- The discipline of the process and people's reactions during the crisis.

- People's duties when the alarm is raised.

- The type and content of the alarm message; it must be short, clear and unambiguous.

9.4.7 Fire development

Facilities building design has a major role to play in fire development. A properly designed building will help prevent the quick development of fire.

Building design

Purpose grouping

Purpose grouping simply means grouping similar activities in the same place, and grouping materials that need similar precautions in the same place, so these precautions can be implemented properly and monitored regularly. Explosive materials and liquids are two good examples. When storing nitric acid for batteries, use a special room where the temperature does not become high. The battery room in telephone exchanges must be situated away from the emergency generator for instance, it must be well ventilated and it must have a smoke detector. Grouping is an important safeguard against fire and it is also economical because these special precautions will only be applied to a part of the location or stores.

Space separation

Building shape is an important way to safeguard against fire. Tall buildings help the spread of fire within the building, due to the effect of wind (Figure 9.8a). The airflow pattern around another lone building is shown in Figure 9.8b. This building is designed with natural ventilation for smoke control, taking account of the prevailing wind flow patterns. When it is surrounded by other buildings (Figure 9.8c) the airflow pattern changes, turning it into a fire hazard. By changing the orientation of the surrounding buildings, the fire hazard can be reduced (Figure 9.8d).

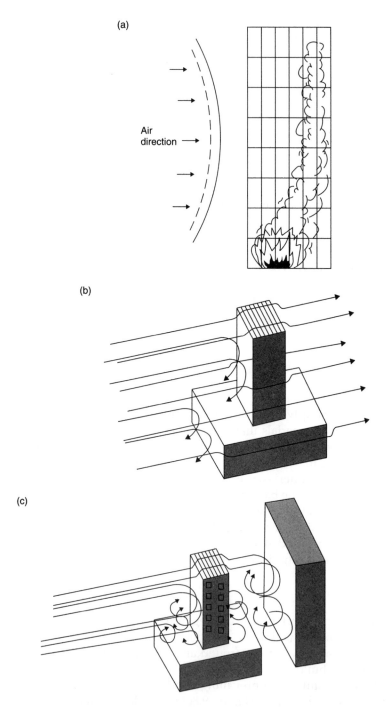

Figure 9.8 Flow patterns: (a) smoke dispersion, (b) airflow round one building, (c) airflow round two buildings, (d) airflow after reorientation

(d)

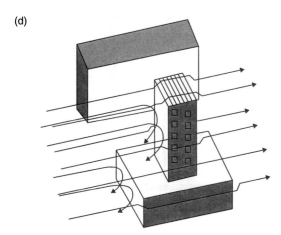

Figure 9.8   (*continued*)

Ignition prevention

Ignition in telecommunications systems can happen during the operation of the standby generator—a well-known cause. Yet it can also happen for other reasons:

- High-voltage and radio frequency discharges

- Sparks from high-capacity capacitors

- Sparking from capacitive pickup cable cores

- Chemical ignition from sodium vapor lamps

- Static discharges from plastic materials

- Sparks in high-voltage cage motors

- Sparks from the commutation process

All these factors (and others, if relevant) must be taken into account when preparing the fire protection plan.

Compartmentation

Compartmentation means performing each of the organization's activities in a separate place, i.e. reject the open-space concept where all the activities of the organization are conducted in a big hall with plywood separators. Fire spreads rather quickly in a big hall and the plywood separators increase the rate of spread and the amount of loss (Figure 9.9).

Each compartment must have an opening in the ceiling or roof, or a window that can be used to vent the smoke. In remote rural exchanges the roof is sometimes made of a material that breaks down quickly under

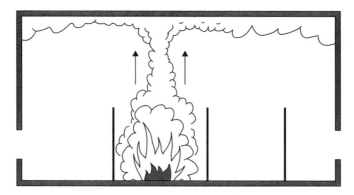

Figure 9.9   Compartmentation

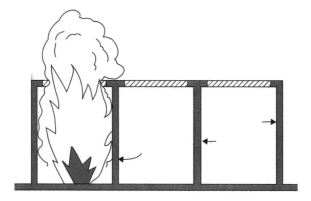

Figure 9.10   Early roof failure

the effect of fire; this insures the fire is confined to only one compartment (Figure 9.10).

The escape route

In each fire protection plan the escape route for the occupants must be studied well to safeguard the lives of the occupants in case fire erupts and smoke develops. The wind direction (outside the building) may sometimes produce positive or negative pressure. If the wind pressure is positive, it will force the smoke inside the building, producing a turbulent mixture of air and smoke, and the escape route becomes smoke clogged rather quickly, preventing the occupants from leaving the building to safety in good time. Figure 9.11 indicates the effect of positive and negative wind pressure on the efficient venting of smoke.

The most important considerations in escape route design are:

- To insure the availability of adequate escape facilities.
- To keep the escape route open and clear of obstacles at all times.

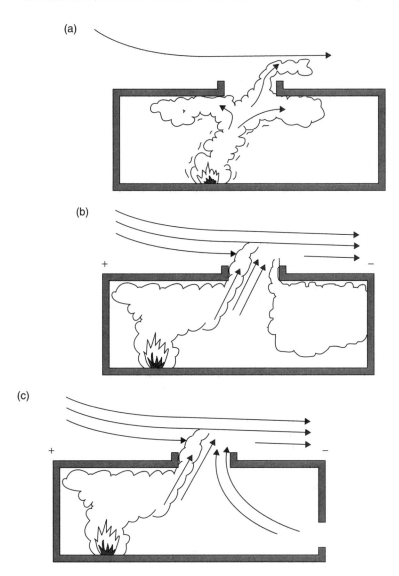

Figure 9.11   Wind pressure: (a) positive, (b) negative, (c) remedy for negative pressure

A well-designed escape route will insure the safety of the occupants and a reduction in overall losses.

Isolation

Fire reaches neighboring buildings in three ways:

- Through dividing walls (adjacent buildings)

- From facade to facade (face-to-face buildings)
- Through the roofs

To prevent fire from spreading through neighboring buildings, use the following protective measures to isolate each building from the other:

- Use dividing walls with a certain level of fire resistance.
- Insure a minimum distance between buildings.
- Enforce strict constraints on the facades.
- Insure the roofing material possesses minimal combustion properties.

Segregation

If we accept that fire in a building is sometimes inevitable, the building can be designed and subdivided so that the damage due to the fire does not exceed a predetermined value. The separating walls in this case are made from material that can withstand a complete burnout. The fire will be confined to the compartment where it originated. The roof should be designed as a fire relief vent. It either melts or burns out in the early stages of the fire, and then provides a good fire vent.

Component integrity

The pillars, girders, walls and other construction elements of the building contribute to the stability of that building. In the event of fire, the building design must guarantee this stability until all occupants have been evacuated.

People evacuation

Statistics show that lack of escape is the most important factor in increasing fire fatalities. This can be attributed to:

- Delayed awareness of the fire
- Escape routes enveloped in fire smoke
- Occupants unaware of an alternative escape
- Unusable exits (blocked, locked or barred)
- More occupants than escape capacity

The risk of injury is increased when people act irrationally. All occupants should be given proper training during fire drills and they should be required to take the training seriously.

The evacuation of the occupants to safety must be planned properly:

1. The occupants must activate the alarm immediately when a fire is identified, in case it is not activated automatically.

2. The occupants must be well trained during fire drills on how to behave and act rationally.

3. Escape routes should be designed to cater for the number of occupants and the estimated number of visitors.

4. Escape routes should be kept clear, unobstructed at all times, and free from smoke during a fire.

5. Primary and alternative escape routes should be clearly marked with large signs.

Time is also an important factor in the evacuation process; it can be divided into:

- The time taken to convince people to leave the premises, especially if they are nonoperational occupants of the building. The longer it takes to convince people to leave, the more fatalities there will be.

- The time to complete the evacuation; this depends on the dimensions of the doors, staircases and corridors of the escape route. As an example, the speed with which the occupants pass through a door 80 cm wide will be 20 m/min; if it is followed by an 80 m corridor of the same width, their speed drops to 10 m/min. Thus a bottleneck is created.

The fire development model

Figure 9.12 illustrates the stages in the development of a fire and it helps to show the activities associated with the firefighting process.

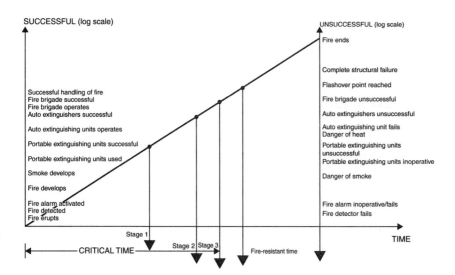

Figure 9.12 Fire development model

Automatic firefighting systems

Automatic water extinguishers are seldom used in telecommunications organizations — water is unsuitable for electrical fires — but gases such as carbon dioxide, halon 1301 or halon 1211 are used. Automatic firefighting systems are designed to detect the eruption of fire, to contain it and then to extinguish it. Gases are used to protect a specific area or areas, unlike water sprinklers, which are designed to protect a whole building. Gas installations can be localized, centralized or modular.

Effectiveness of fixed installations

To illustrate their effectiveness, statistics show that, using fixed firefighting installations, 95–99% of all fires were extinguished before the fire brigade arrived. The remaining 1–5% were extinguished after the fire brigade arrived

Pros and cons of firefighting materials

Carbon dioxide

✓ Has the lowest cost

✓ Leaves no residues

✓ Safe with electricity

✓ Local use or on all premises

✓ Works on most fires

✓ Okay on deep-rooted fires

× Chills air, condensation

× Impairs visibility

Halon gas

✓ Works on most fires

✓ Unimpaired visibility

✓ Safe with electricity

✓ Leaves no resides

× Expensive to recharge

× Not okay on deep-rooted fires
    — concentrations >10% are hazardous

× Escapes through openings
    — escaping gas lowers the efficiency

Dry chemical

✓ Works on most fires

✓ Safe with electricity

✓ Quick knockout

✓ Puts out metal fires

× Impairs visibility

× Leaves residues

× Needs cleanup

× Transitory effects

Firefighting by people in the building

Time is vital in firefighting; early detection and fighting help to contain and put out the fire quickly. Telling occupants about fire-handling techniques and training them not to waste any time in coping with fire, will help in saving the organization's buildings from a disaster when fire erupts.

What the occupants should do

Occupants should *not* observe fire, report it to the fire service and then wait for help. On the contrary, they must take an active role as soon as they identify a fire, according to what they were trained to do during fire drills. Until the fire brigade arrive, occupants have a very important role in helping to fight and contain the fire, and prevent it from spreading to other locations.

Why the occupants' contribution matters

• They are the first to notice the fire.

• They can guide the fire brigade and other occupants to access points and escape routes.

• They can locate the electricity main switch to switch off electricity.

• They can furnish vital information about the other occupants in the building.

• They can furnish vital information on the appropriate choice of extinguisher.

How the occupants should behave

The occupants can behave either rationally or irrationally. Usually when the occupants are well trained, they will behave rationally. Their reaction will be irrational if they find themselves facing a danger from which they think there is no escape. Here are two examples:

- The smoke and fumes may develop quickly and begin to influence the occupants' ability to react rationally. Fumes are opaque, toxic and irritant; caught in the fumes and cut off from the outside world, the occupants may panic.

- When the escape route is long, congested or complex, it gives the occupants the bad feeling that they will not make it to safety in time; they begin to panic.

The importance of fire drills

Fire drills are very important. They train the occupants to handle fire situations rationally, helping them not to panic.

Here are some important points when conducting a fire drill:

- A drill should be made at irregular intervals and without prior announcement.

- Each occupant should know at least two different escape routes from their workplace, and they should remember the nearest location of the fire extinguishers.

- Each occupant should know the proper way to operate the fire extinguishers.

I once worked for a company that made TV tubes. The glass tubes had to be heated in furnaces to relieve any stress, and the furnaces were fueled by large containers of butane gas. When analyzing the hazards in the factory, we identified that a disaster could happen if one of these containers caught fire and blasted. During fire drills we asked the fire-fighters to demonstrate to the factory workers how to stop a container from blasting when it is on fire. A practical demonstration was made and each worker got the chance to turn off the valve of a burning container (using protective gloves). This was a unique experience for the workers, who had never thought it possible until then. This drill helped save the factory when a container caught fire by accident.

Fire drills are also important in indicating the actual efficiency of the escape routes, and in testing their condition. At times some of these escape routes become clogged, or their doors are locked and the keys are not in their proper place. If a fire occurs under these circumstances, the penalty will be high.

Firefighting equipment at hand

The occupants of each building should be familiar with the firefighting equipment in the building and they should be able to use it effectively.

Portable extinguishers

The most important factor in the construction and design of portable extinguishers is that they can be used and operated by everyone. In fact, the operating procedures should be straightforward and presented in a pictorial form, so that everyone can learn without reading the instructions. It is better to use extinguishers with the operating instructions drawn on the barrel, then language is not a problem; this is especially important in developing countries where many people cannot read.

Hose reels with axial water supply

Water can be used when no electrical connections are under fire. Effective firefighting depends on the condition and length of the water hose.

Fixed extinguisher installations

Fixed extinguisher installations are available in organizations at specific locations. There can be several types of extinguisher in any organization, depending on the type of fire expected. It is most important that people choose the right extinguisher to tackle each fire.

Escape from fire

Escape route design

Here are the most important considerations in designing an escape route:

- Insure the escape facilities are adequate for the number of occupants that will use them when fire erupts.
- Insure the escape route is available, well marked, and clear of obstacles at all times.

Escape route planning

The aim of route planning is to insure the occupants, from all parts of the building, will be able to reach a safe place within the building, or better outside the building, and in good time. In calculating this time, consider the average speed of a walking person (or running person) and the length of the escape route itself. Studies have shown that the speed with which people run in an open space can be very quick, but in rooms and corridors we must consider the space restrictions. Four cases can be treated for occupants having different speeds and in different situations:

- Normal healthy people can walk quickly inside the building at a speed of say 30 m/min.
- People walk normally inside buildings at a speed of say 12 m/min.
- Old and disabled people walk at a speed of say 5 m/min.
- Old and disabled people on a congested escape route may walk at only 2 m/min or even less than that.

Fire protection zones

When planning the escape route, divide it into three zones (Figure 9.13):

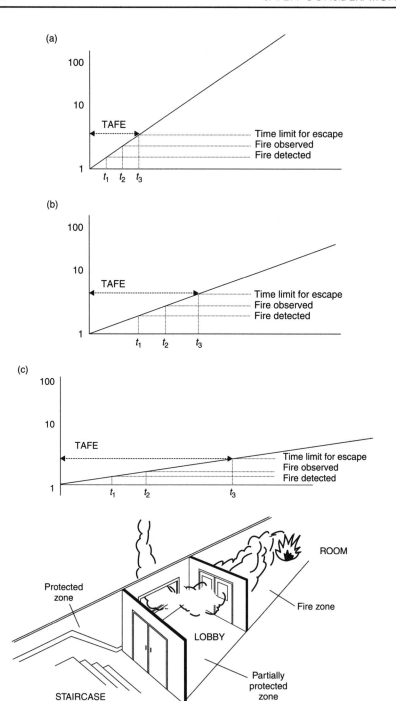

Figure 9.13 Fire spread: (a) in the fire zone, (b) in the partially protected zone, (c) in the fully protected zone, (d) view of all three zones

Table 9.2 Escape route speeds

Condition	Escape speed (m/min)
Escape route congested	1–5
Escape route normal	
Disabled people escaping fire	1–5
Normal people walking from fire	10–12
Normal people running from fire	10–25

* *The fire zone*: where the fire has erupted.

* *The semiprotected zone*: the occupants use this zone to reach safety; it must have some protection from the fire otherwise the occupants will be trapped and unable to reach safety.

* *The fully protected zone*: the occupants use this zone to reach safety.

Estimate the time needed to evacuate each of the three zones; this estimate should consider the possible types of fire, their rates of spread and the speed of the occupants in each zone. The safety route length is a critical design factor (Table 9.2). The data in Table 9.2 and the speed of escape for each zone are explained further in Figure 9.13. Notice that the time available for escape increases with increasing protection against fire spread. This time is indicated on the curves as TAFE (time available for escape).

Warning and alarm systems

Any alarm system is supposed to alert the employees in the telecommunications company to the presence of a life-threatening event. In all cases the contents of the message should be clear and unambiguous. The selection of an alarm will depend on several factors:

- The population matrix of the employees, their types, perception and decision-making ability.
- The population discipline, how they will behave when they hear an alarm, and what kind of action they are going to take.
- The workplace (telecommunications company, TV station, computer facility, etc.); each type of work will dictate a certain type of alarm system.
- The nature of the message; is it for information only or does the recipient have to act on it?
- How the alarm is raised: aural, visual or a combination of both.

Protective clothing for firefighters

Firefighters wear special fireproof clothing to protect them from flames and high temperatures. They must also wear smoke masks so they can rescue people trapped in smoke-filled areas.

Table 9.3 Hazard classification

UN class	Hazard type	Hazard details
1	Explosives	Mass explosion hazard
2	Gases	Compressed, liquefied or dissolved under pressure
3	Flammable liquids	Flash point
4	Flammable solids	Spontaneously combustible substances Substances giving off flammable gases in contact with water
5	Oxidizing substances	Substances other than organic peroxides Organic peroxides
6	Poisonous substances	Infectious substances
7	Radioactive substances	Can cause cancer
8	Corrosive substances	
9	Miscellaneous dangerous substances	

9.5   MANAGEMENT OF DANGEROUS MATERIALS

9.5.1   Primary hazards

Hazard classification

It is difficult to define a hazardous material, because many substances, when transported by air, sea, rail or air, can prove dangerous or hazardous. Yet we use them in performing our daily work duties, taking the necessary precautions. But the danger of hazard depends on the degree of toxicity, corrosiveness, explosiveness or flammability of the substances (Table 9.3).

Hazard identification

Each hazard has a diamond-shaped label with a certain color according to the color identification in Table 9.4.

9.5.2   Handling hazardous materials

The most important hazard in the telecommunications industry is in storing or transporting (by road) hazardous substances (fuel). Acid spill during battery maintenance is another hazard.

Table 9.4 Hazard identification

Hazard type	Hazard identification color
Explosive	Orange
Flammable	Red
Water reactive	Blue
Oxidizer	Yellow
Toxic/infectious	White
Corrosive	Black and white

Road transportation

Moving substances by road is not a hazard in itself, but hazard occurs if there is an accident on the road (or malfunction of the car), resulting in spills or fire. These effects can be minimized if some precautions are taken:

- The vehicle is tested before the trip to insure it is roadworthy.

- The substance is well arranged and securely packaged.

- The package is compatible with its hazard classification.

- The package contains only one hazardous substance; it is dangerous to pack a hazardous substance with other noncompatible hazardous substances.

Movement precautions

- The vehicle carrying hazardous substances should be directed away from areas of high risk or tunnels.

- The vehicle should be clearly marked to indicate it is carrying hazardous substances.

- The vehicle team should be properly informed on how to deal with rescue and recovery during an accident.

Route selection

- *Risk assessment*: to assess the risks along the vehicle's route.

- *Probability assessment*: to assess the probability of an accident happening to the vehicle; this will vary with the type of substance and the condition of the road.

- *Cost viability assessment*: to assess the commercial viability of the selected route — a good toll road versus a bad freeway.

- *Accommodation assessment*: to assess the availability of emergency services and rescue personnel.

(a)

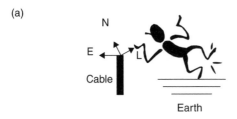

Earth

(b)

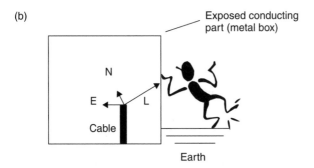

Earth

Figure 9.14   Electric shock: (a) direct, (b) indirect

9.6   ELECTRICAL HAZARD CONTROL

9.6.1   The philosophies of earthing

The importance of earthing stems from the fact that if a voltage of 50 V or more touches, or is applied across, a human organ, blood may cease to flow to the brain and this may eventually cause death by electrocution (Figure 9.14). The International Electrotechnical Commission IP (ingress protection) code differentiates between two types of equipment:

- *Class I equipment* has exposed conducting parts.
- *Class II equipment* has double-insulated, unearthed enclosures (e.g. the electrical water-heating kettle).

9.6.2   The dangers of improper earthing

Single-phase circuits

Single-phase class I equipment can have a faulty earth in three ways:

- The live conductors touch the metal casing and the casing is not earthed. The metal enclosure becomes live, and anyone touching it will receive an electric shock.
- The live conductors touch the metal casing and the casing is earthed. The metal enclosure becomes live and immediately enables a fault

current to blow the fuse. Anyone touching the enclosure will be safe; they will not receive an electric shock.

- The neutral conductors touch the metal casing and the casing is earthed. The neutral fault to earth remains undetected, but anyone touching the enclosure will not receive an electric shock.

Three-phase circuits

The general rules of single-phase circuits (regarding electric shock) apply for three-phase circuits. There are several earthing systems for three-phase circuits, and each one has particular characteristics which should be studied carefully by the maintenance engineer:

- *The TN-C system*: here the neutral and protective conductors are combined together throughout the system, so we have four wires running from the energy source to the consumer. All exposed conductive parts of the installation are connected to the combined protective and neutral conductor (PEN).

- *The TN-S system*: separate neutral and protective conductors are used throughout the system, so we have five wires running from the energy source to the consumer. All exposed conductive parts of the installation are connected to the protective conductor (PE), which is the metallic covering on the cable supplying the installation, or a separate conductor.

- *The TN-C-S system*: here the neutral and protective conductors are combined together as a single conductor in a part of the installation, so we have four wires running from the energy source to the consumer, but we have five at the consumer's installation. This system is also known as protective multiple earthing, and the PEN conductor is known as the combined neutral and earth (CNE) conductor. All exposed conductive parts of the installation are connected to the PEN conductor via the main earthing terminal and the neutral terminal, which are linked together.

- *The TT system*: this has one point of the energy source directly earthed; the exposed conductive parts of the installation are connected to earth electrodes electrically independent of the earth electrode of the source. Thus we have four wires running from the energy source to the consumer (three lines and one neutral).

- *The IT system*: this has no direct connection between live parts and earth; the exposed conductive parts of the installation are connected to an earth electrode. The source is either connected to earth through a deliberately introduced earthing impedance or it is isolated from earth. Thus we have three wires running from the energy source to the consumer (three lines).

9.6.3 Cable fires

All cable insulating material can be assessed and evaluated according to the following important parameters:

- *Oxygen index* is the minimum percentage of oxygen in a mixture of oxygen and nitrogen, which will sustain combustion. When this amount is less than the proportion of oxygen in the atmosphere (about 21%) then the cable insulating material will burn freely in air.

- *Hydrochloride emission* is the percentage by weight of hydrochloric acid emitted during the combustion process.

- *Smoke emission* is the density of the smoke emitted after the material has been subjected to a standard flame in a standard test cell.

- *Self-ignition temperature* is the temperature at which the cable insulating material will start to ignite.

Cable fires are rarely caused by failure of the undisturbed cables themselves, because they are always protected by fuses and circuit breakers. Fire in fact originates from the quality of the organic combustible material in the crowded busy cable runs. That is why cables must have minimal emission of smoke, toxic fumes and acid gas, under fire conditions.

Current rating: factors to consider

Maximum ambient temperature

A correction factor C_a is applied to the tabulated current-carrying capacity, depending on the actual ambient temperature of the location in which the cable is installed. The correction factor is 1.0 at 30°C but reduces as the actual ambient temperature increases. For PVC cables the temperature correction factor reduces to 0.87 at a temperature of 40°C, reaching 0.35 at 65°C. Thus, at an ambient temperature of 65°C, current-carrying capacity is taken as one-third of the tabulated current-carrying capacity.

Number of cables bunched together

Where cables or circuits are grouped, a correction factor C_g is applied to the tabulated current-carrying capacity. The correction factor is 1.0 for a single cable but drops to 0.65 if four cables in conduit are bunched and clipped direct to a nonmetallic surface.

Overload protection

When overload protection is used in the circuit, the operating conditions of the cable are not only affected by the ambient temperature, stacking method and cable grouping, they are also affected by the conductor temperature, which may rise during the period it takes the overload device to operate. Thus a factor is included to cater for the type of overload device in use.

Installation method

A correction factor is applied to the tabulated current-carrying capacity, depending on the cable installation method, since this will affect the thermal insulation of the cable. It is incorporated into the calculation of C_g.

Cables for telecommunications

Polyvinyl chloride cables

Polyvinyl chloride (PVC) is a thermoplastic resin resistant to moisture, dilute acids and alkalis. It is less flammable than rubber, it allows untinned copper conductors to be used in cables. It has a relatively high dielectric loss, so it is rarely used for AC voltages over 6.6 kV. It is designed to work at a maximum temperature of 70°C, but improved types can withstand 100°C.

When flame is applied then removed from the cable, combustion of the insulation does not persist, thus PVC is self-extinguishing. Yet PVC cables can propagate fire if ignited. When PVC burns it generates about 25% by wt HCl gas, which has a corrosive and choking effect; it also emits opaque black smoke even before the insulation catches fire. Smoke blocks all light and obscures the exit signs and the escape route.

Mineral-insulated cables

Mineral-insulated copper-sheathed cables (MICCs) have better refractory properties, making them a prime choice for fire alarm systems, emergency lighting systems and other circuits which must continue to function during and after a fire.

The surface temperature of the cable has to be restricted to 70°C, since above 250°C the copper begins to oxidize, but the cable continues to function properly at temperatures of about 1000°C for a short period; this period must be sufficient for the alarm system to give the needed fire warning signals.

The main disadvantage of MICC is that the magnesium oxide is hygroscopic, and unless sealed from the atmosphere at each termination, the insulation slowly turns to the hydroxide, which is a conductor, rendering the cable useless within the control system.

Cross-linked polyethylene cables

Polyethylene cables are better than PVC because they can work at −60°C and they have a lower dielectric loss factor, which makes them more suitable for high voltages. Polyethylene cables contain no halogens so they burn without producing corrosive gases.

They can be used up to a temperature of 90°C, valuable for insulation in hot climates. Cross-linked polyethylene permits tighter bends than rubber and PVC, and it does not require vapor-proof sheathing with sealed joints and terminal boxes, so its insulation is cheaper and its welding is simpler.

9.6.4 Electrical equipment in explosive atmospheres

The widespread use of natural gas and petroleum means that a major hazard is accidental ignition of flammable gases and vapors. Some developing countries are still using gas burners to weld lead-sheath telecommunications cables in access holes. The hazard situation develops when gas escapes unburned and mixes with air to form a large volume of explosive mixture. On ignition the energy released will create a sudden increase in temperature, before the mixture has time to expand or lose heat to the surroundings.

When this happens in the access hole, the rise in temperature causes the pressure to increase severalfold, destroying the access hole and perhaps killing some of the technicians. The same situation can happen in mines, where gas mixes with combustible dusts to cause highly destructive explosions. It is extremely important to prevent uncontrolled or undetected escape of flammable gas or vapor. Here are the sources of electrical ignition responsible for such a disaster:

- Sparks produced in normally nonsparking equipment due to internal circulating current (cage motors).
- Static discharge from plastic covers and enclosures.
- High-voltage and radio frequency discharges.
- Sparking from inductive and capacitive pickup or unused cable cores.
- Chemical ignition from discharged sodium vapor lamps.

9.7 PROTECTION FROM LIGHTNING STRIKES

9.7.1 The development of lightning

Lightning is a transfer of electrical charge. A thundercloud has a tripolar separation: a main central negative charge with positive charges above and below. When the voltage between two oppositely charged fields exceeds atmospheric resistance (30 000 V) then lightning results. This may occur within a cloud, between clouds or between clouds and earth.

When there is a lightning strike from a cloud to earth, a negative current is carried downwards. Within 100 m of the ground, a return positive charge moves upwards. The downward strike is slower, so the observer perceives the strike to be from the cloud to the ground. Visible flashes average 5.5–6.0 m in diameter and 1.6 km in length.

9.7.2 Types of lightning strike

- *Direct strike* hits the victim directly and most often involves persons in the open or those in contact with metal objects

- *Splash* occurs when lightning strikes an object and jumps to a nearby person of lower resistance. This is the most frequent mechanism for lightning injuries.

- *Ground current* occurs when a bolt hits the ground and is transferred to a person. Injuries from ground current are inversely proportional to distance from the strike.

- *Blunt trauma* injuries result from shock waves due to the expansion or explosion of rapidly cooling superheated air. This may cause blunt trauma directly or indirectly through collision injuries. Blunt injuries occur in 32% of lightning victims.

- *To summarize*, splash and ground currents hit most persons injured in multiple-victim strikes; the most severe injuries are the result of direct strikes.

9.7.3 Injuries from lightning

Injury from a lightning strike may be caused by three types of energy: electrical, thermal and mechanical. Electrical energy damages the tissues in several ways. Tissues that rely on electrical conduction, especially the heart, can be electrically discharged and require recovery before resuming function. The extent of electrical energy is determined by voltage, amperage, resistance, anatomic pathway, duration of current and type of circuit.

Injury types

- *Central nervous system*: the central nervous system is quite sensitive to lightning injury, which may present as a period of impaired consciousness, transient disorientation and confusion, but some victims have seizures and varying degrees of paralysis.

- *Skin*: most lightning victims sustain burns, usually first- and second-degree burns. Third-degree burns may occur on the skin under metal objects such as jewelry, or at the lightning entry and exit points on the body.

- *Heart*: about 50% of lightning victims have nonspecific cardiac dysfunction.

- *Eye*: lightning strikes cause many ocular injuries, including optic nerve atrophy. Approximately half of all lightning victims exhibit some eye damage, the majority having corneal injuries and cataracts.

- *Ear*: ear injuries include rupture of the tympanic membrane and middle ear injuries.

- *Psychiatric symptoms*: psychiatric reactions to lightning strikes are similar to those observed in disaster studies; they include storm apprehension, bad dreams and severe depression. Children are more prone to psychiatric reactions; they present with anxiety, sleep disturbances and nightmares.

9.7.4 Lightning survival techniques

Lightning kills and injures many people each year. Important differences exist between lightning victims and patients injured by fire or by other forms of electricity. Most lightning fatalities and injuries occur during the rainy season. Death occurs in 20–30% of lightning victims and is usually due to cardiac or respiratory arrest. At the scene of a lightning strike, living victims will almost always survive, but those who appear dead (in arrest) need assistance. Resuscitation attempts have a higher success rate in lightning victims than in persons with other types of cardiopulmonary arrest, even among patients with conventional signs of brain death. An overall resuscitation success rate of 50% has been reported.

9.7.5 General guidelines for survival

- Do not camp near tall trees, next to bodies of water or on the highest hill in an area. Always be on the alert for a thunderstorm.

- The safest place during a thunderstorm is inside. Since lightning usually passes along the surface of conductors rather than through them, people in buildings and cars are rarely injured by lightning. In buildings, do not stand between an open door and an open window, since lightning may travel between them. Avoid telephones during thunderstorms.

- If caught outside during a thunderstorm, do not seek shelter under any structures that present the shortest path for lightning (i.e. any isolated tall objects such as a tree or a tent). These serve as lightning rods, and people sheltered under them are at higher risk for both direct strikes and splash strikes. If the vicinity of trees cannot be avoided, keep clear of the tree by at least 3 m. While facing the tree, keep the feet and knees as close as possible to avoid the buildup of a step voltage across the legs. Statistics show that people get killed more frequently under a tree than out in the open.

- If caught in an open area, get away from isolated machines, such as tractors and golf carts. Discard fishing rods, golf clubs, bicycles and umbrellas. Get away from bodies of water. Head for dense woods or lie on the ground in a ditch, with a rubber poncho or raincoat underneath to decrease electrical grounding. You may also kneel down and bend forward with hands on knees; this presents as small an area as possible for lightning to strike.

9.7.6 Management of lightning victims

Make sure you understand the triage priorities at the scene of a lightning strike. Delay treatment of the moving, moaning victims and instead resuscitate the apparently dead victims. The goal is to oxygenate the heart and brain until the heart regains its electrical potential and respiratory depression passes. Resuscitation efforts should continue longer in lightning victims than in victims of other trauma.

9.8 PERSONNEL SAFETY CONSIDERATIONS

9.8.1 Protecting vulnerable personnel

Human interactions during a disaster

The purpose of the analysis is to find the human interaction in incidents and to identify any negative human influence. Human errors can be classified into the following categories:

- Erroneous operations
  - Erroneous operation of the equipment
  - Bad alignment
  - Reversal of cables, etc.
  - Omission or inhibition of certain functions
  - Programming errors
  - Improper labeling

- Operating errors
  - Maloperation of the equipment
  - Errors in the operating procedures
  - Misinterpretation of results
  - External events

- Maintenance and test errors
  - Omission of maintenance
  - Improper implementation
  - Improper calibration
  - Fabrication defects

- Design errors

Causal analysis of accidents

Work organization

- Actual content of the work and how well it is understood
- Level of monotony in the work; the more monotonous the work, the higher the probability of an accident
- Possibility of a conflict between the operational procedures and safety
- Level of initiative required to perform the task
- Information available on the task; if information is scarce, the probability of an accident will be high

Design of the work situation

- Arrangement of indicators with respect to the operator's visual field
- Design and readability of the instrumentation
- Readability of the instructional procedures
- Design of the sound and visual alarm
- Restrictions imposed by safety equipment
- Compliance with existing color standard codes

Time and duration of the work

- Duration and variability of the work timetable
- Period of work
  — shifts
  — night
  — permanent
  — seasonal
  — overtime
- Duration of breaks
- Working after an unscheduled outage

Personnel and individual factors

- Period of general training
- Period of professional training
- Training method used to perform the task
- Number of years spent at the same work
- Number of years in the organization

Physical environment

- Level of noise at the workplace
- Amount of vibration during the work
- Workplace lighting and color schemes
- Radiation exposure of each worker
- Amount of smoke, dust and toxic products

Social environment

- Family life of the workers
- Stability of the employer
- Company turnover
- Conflicts in the organization
- Attitude towards other employees
- Performance stimulus and incentives
- How much they tell the organization

History of the plant

- A plant that has been running without accidents creates a feeling of tranquillity within the employees
- A plant that has been running with a lot of incidents, outages and accidents creates uneasiness among the employees and leads them to feel that something is always about to happen

Employee characteristics

- Age, nationality, sex, marital status
- Degree of dependability and self-assurance
- Stress-handling capacity
- Ability to concentrate at work
- Hearing, vision and color blindness

Procedures

Procedures are the rules that interface between people and the system. There are three major areas to consider:

- The technical content of the procedures
- How the technical procedure is presented:

— Adapt it to each operator's level of education and training
— Show the actions taking place at the same time, the allowable and unallowable steps, and the role of each operator in the team
— Indicate clearly what actions should be taken in case of failure, or an outage
— Train the operators adequately on how to perform the tasks detailed in the procedures

• The potential errors during implementation

Accident proneness

Statistics show that a small minority of workers encounter more accidents than can be explained by chance alone (normal distribution). An individual might have some personality trait which predisposes them to accidents. In some people, accident proneness is a passing phase, while in others it is more enduring. This is the fault of the man–machine system with all its complexities. If such workers are identified, they should be relocated to places where the probability of fatal accidents is remote.

9.8.2 Ergonomics

Introduction

The term 'ergonomics' was coined from the Greek *ergon* 'work', and *nomos* 'natural laws'. It is a discipline which attempts to redress the balance between human beings and their environment. This balance has become increasingly important since the Industrial Revolution, particularly with the expansion in the complexity of both work and machines. Because of a poor fit between the human operator and their environment, lives have been lost, productivity reduced, and errors have been incurred in countless thousands of cases. Until relatively recently, the demands of the work environment have been paramount, with the workers themselves taking second place. But there are limitations to this approach:

• It may be costly to fit the operators to their environment

• It may not be effective

• Operators may suffer stress and this could disrupt their performance

Training the operator for tasks which are difficult to carry out is a costly procedure. Although training schedules cannot be dispensed with altogether, experience has shown that, in the majority of cases, training and production time can be reduced considerably if the machine is designed to reflect the operator's abilities. It is a fact that no amount of training will overcome the tendency of an operator to do what comes naturally under stress, except in some military training programs.

The scope of ergonomics

- *Physiology, anatomy and medicine* provide information about the structure of the body, the operator's physical capabilities and limitations, the dimensions of the body, how much they can lift, the physical pressure they can endure, etc.

- *Physiology and experimental psychology* provide information about the functioning of the brain and the nervous system as they determine behavior, how humans use their bodies to behave, to perceive, to learn, to remember, to control motor processes, etc.

- *Engineering and physics* provide information about the machine and the environment with which the operator has to contend.

9.8.3 First aid techniques

It is important that all employees, let alone the fire brigade, should be well informed on the basic rules of first aid. Yet the first important rule is to learn to be fit. If you are fit, you can act properly in a disaster; this includes the following items:

- Exercise and know why exercise is good for you

- Relieve tension regularly

- Get a good night's sleep

- Detect stress whenever you feel uncomfortable

- Avoid drinking and smoking

- Implement the safety rules and regulations

- Identify your occupational and environmental risks

- Know how to handle casualties until the first aid people arrive:
  — absence of breathing
  — unconsciousness
  — absence of heartbeat
  — severe bleeding
  — electric shock
  — minor and severe burns
  — fractures and dislocations
  — sprains and strains

9.9 ENVIRONMENTAL HAZARDS

Telecommunications equipment and systems are affected by several environmental hazards that might cause disaster:

- High levels of humidity in the atmosphere
- Corrosive elements in the atmosphere
- Mechanical vibrations and shocks during operation
- Exposure to X-rays and γ-rays
- Attacks by insects and fungi
- Severe variations in pressure
- Extreme variations in temperature

Consider how these hazards affect the system; make it a priority when preparing the disaster management plan. Devise methods to reduce their effects and implement appropriate remedies.

SUMMARY

Safety is an important aspect of disaster management; it helps to prevent a disaster by eliminating the causes.

A safety system is the set of procedures, staff and equipment specifically designed to be applied by the organization to increase its safety. To insure a safe system, each telecommunications organization has developed and formulated a set of design guidelines to be followed in the manufacture, installation, testing, operation and maintenance of telecommunications equipment. These guidelines are concerned mainly with the safety of the equipment, the employees and the customers. Engineering safety helps increase the company's profits by reducing the incidence of failures and the associated loss involved.

Firefighting, control and prevention were discussed in some detail since fire is one of the major disasters that affect telecommunications organizations. Dangerous materials can cause disasters if some safety precautions are not followed. Electrical equipment must be properly earthed. Cable fires are important hazards. Protect buildings and people against lightning strikes. Lightning victims need to be treated differently than other trauma victims. If a worker is accident prone, find them a job where the risks are minimal.

REVIEW QUESTIONS

1. Define the following terms: a safe system, the Wiring Regulations, UL listing, extinguishing mechanisms, active fire control, passive fire control, vulnerable personnel, accident proneness, environmental risks.

2. Describe how you would protect your workplace from the dangers of fire.

3. Compare the different types of extinguisher and select a system suitable for your workplace. Indicate the reasons for your choice.

4. Describe how you would protect your workplace against lightning. Select a system for your workplace. Indicate the reasons for your choice.

5. Describe the possible injuries to employees during a lightning strike.

6. Describe how you should manage lightning victims.

7. How can you identify accident-prone personnel? Describe how you should manage them at your workplace.

10

Legal Issues in Disaster Management

OBJECTIVES

- Answer questions at the end of the chapter.
- Discuss the legal issues in disaster management.
- Present some legal problems at your workplace due to service interruptions.
- Describe the legal aspects of service interruption disasters, and their effects on your company's liabilities as a service provider or as a customer.
- Say how you propose to handle the legal issues of service interruption at your workplace.

10.1 INTRODUCTION

Telecommunications services have grown almost exponentially in the past few years, transforming the world into a small village, very well interconnected by all means of telecommunications and broadcasting facilities. The introduction of the Internet has transformed the world into a huge market, and electronic commerce is growing at an enormous pace. This seems very good from the viewpoint of telecommunications investors, but it is very alarming from the viewpoint of telecommunications managers and O&M engineers.

This alarm is justified if one considers the cost of service interruptions. That is why contingency and service recovery planning has become a way of life in the telecommunications industry. Almost all businesses are now using computers to conduct their work, and telecommunications data transmission media to exchange information with other sections of the business, or with other affiliated enterprises.

The cost of interruption is excessive in the case of large-capacity transmission media, such as satellite links. One aspect we have yet to address
is the cost arising from the effects of interruption on the customer. This
cost is termed liability and it may be very large.

10.2   LIABILITY

When the telecommunications organization makes an agreement with
another organization that uses the lines to transmit information, the
agreement usually stipulates what should be done in case of a disaster,
i.e. when the telecommunications organization cannot provide the service
for a long or short time. This kind of agreement usually addresses the
following points:

- The organization's duty to insure continued service to its customers.
- The limits of this duty, how and when it can be held liable to the losses
  that may be suffered by its clients due to the interruption of the service.
- The duty of the O&M staff to insure continued service.
- The limits of this duty, how and when the managing director or even
  the O&M staff can be held liable for the losses that may be suffered
  by the organization's clients due to the service interruption.
- The formula that will be used to calculate any compensation.

10.2.1   Private companies

When a private telecommunications company fails to provide the
continued service as contracted, business customers may suffer losses due
to that reason. They may be tempted to recover all or part of their losses
from the telecommunications company, since liability can be imposed
on a private company under statutes and under common law. There are
several legal actions that can be taken.

Negligence

The plaintiff must establish, beyond any reasonable doubt, that the defendant had a duty to act, in the expected manner in performing his/her
duties, and he/she violated that duty by an act of negligence. The
defendant's duty is established by examining the telecommunications
company's technical procedures, the job description and duties of the
staff involved, their responsibility limits, and the company's authority
delegation procedures.

Professionalism

The plaintiff must establish, beyond any reasonable doubt, that the defendant had a duty to act, in a professional manner, and he/she violated that

duty. The plaintiff must establish that the action taken by the defendant would not be taken by a reasonable and prudent person in a similar situation. This may suggest that the professional competency of the defendant should be questioned.

The company directors

The company directors can be held responsible for actions taken by themselves or their subordinates, if it can be proved that the disaster was due to the execution of specific orders given by the directors. This is a very critical point since usually the time factor is involved. Let us consider the following example.

A shift engineer responsible for a megawatt medium-wave transmitting station was alarmed to find that the station went off air during the speech of the president of the republic. He checked the radio frequency final stage and found it okay, he connected the dummy load to the station output and operated the station and found it okay, so he inferred that the problem was either in the 2 km transmission line or in the antenna itself. It was 9 pm and dark outside, so he took the station wagon and drove alongside the transmission line and found it okay. He checked the antenna and found it okay. Checking the matching unit, he found a burnt high-voltage capacitor.

Going back to the station store to get a new one, replace the defective one, and put the station on air again took 15 minutes. The total interruption time was 30 minutes. The engineer was questioned and, after hearing his side of the story, the panel convened for three hours. Then they informed the shift engineer that when he discovered the fault, he should have sent two groups of technicians to help, one to check the transmission line and the other to check the transmitting antenna. He should also have alerted the storekeeper. The investigation panel thought this could have reduced the interruption time to the maximum allowed time of 5 minutes.

The shift engineer's situation was as follows:

- He had only one technician in the shift who was only responsible for the operation, and he had no idea about faultfinding.

- The stores were closed after working hours at 4 pm, and the shift engineer was forced to get the storekeeper from the staff quarters each time a fault happened.

- The panel was composed of three engineers who spent three hours deliberating to come to their decision. The shift engineer was alone and had only a few seconds to think.

- The shift engineer was found guilty.

This case on the liability of private telecommunications companies illustrates the situation under statutes and common law applied in the Anglo-Saxon legal family, as in the United States and the United Kingdom.

But in comparative law, as in the Latin legal system, another rule is applied in this respect. The Egyptian civil code, for example, states that:

> In case of non-performance of the contract, the court shall order the debtor to pay damage to the creditor as compensation, unless the debtor proves that the impossibility of performance arose from a cause beyond his control. The same rule shall apply if the debtor is late in the performance of his obligations. (Article 215)

Accounting standards

The corporation can be held liable, under statutory and common law, if it fails to assure the continued operation of its services to its clients. The corporation is required to comply with the recognized record-keeping procedures as well as having adequate accounting control requirements, so that its customers are sure it will not become bankrupt at any time. Usually any telecommunications organization is required, under law, to insure the availability of:

- Accurate and detailed books, records and accounts
- A system of internal accounting control that will insure the financial operations and assets of the company are adequately controlled

10.2.2 Public entities

Since private telecommunications companies can be sued by their customers when they suffer losses due to a telecommunications disaster, governmental entities might also be sued under similar circumstances, when the disaster renders the telecommunications entity unable to provide the service or when the entity cannot recover normal operation quickly.

Unfortunately, in most developing countries, public telecommunications entities can avoid such liability because of well-established policies which restrict the limits of the government's liability. Certain clauses to that effect are usually incorporated in the contract signed between the public telecommunications entity and the customer. The net result is that the government cannot be sued unless it agrees to be sued. If the government agrees to be sued, it will indicate clearly the extent to which it has consented to be sued.

This is the problem of contract of adhesion. The acceptance in this case is confined to adhesion to standard conditions which are drawn up by the offering party (i.e. the Egyptian Telecommunications Company) and which are not subject to discussion. To resolve this problem, the Egyptian Civil Code states that:

> When a contract of adhesion contains abusive 'Lionin' conditions, the judge may modify these conditions or relieve the adhering party of the obligation to perform these conditions, in accordance with the principles of equity. (Article 149)

10.2.3 Common carriers

When a telecommunications satellite fails to continue operating successfully and the spare satellite does not operate at the right moment (which hasn't happened so far), the business will suffer very much since all the international business transactions will be disrupted. Any common carrier cannot bear the total cost of lost business or the cost of loss claimed by their customers. So there must be a limit to the compensation paid by the telecommunications company.

In May 1988 Illinois Bell's Hinsdale exchange fire disrupted telecommunications services to thousands of customers and had devastating impact on the business. In January 1990 the AT&T switching facilities failed for nine hours and disrupted business operations for its customers. Neither Bell nor AT&T could possibly have borne the full cost of the damage inflicted on its customers. Thus there must be a limit to company liability in such a disaster. In real-life situations, almost all telecommunications companies would be happy to compensate their customers for the tariffs charged over the period of interruption.

10.3 LIMITATIONS TO LIABILITY

The costs of telecommunications service interruptions are huge, and a company cannot compensate its customers for the actual and total loss of business. The service provision contracts that companies sign with their customers stipulate the limits of their liabilities.

Furthermore, it has been ruled in the courts that a telecommunications company's liability may not be limited to the tariffs, when there is gross negligence or willful misconduct. Yet it is very hard to prove gross negligence or willful misconduct in court.

It was further ruled in the courts that it is acceptable to limit the company's liability to third persons that were affected because the customer's telecommunications services were disrupted. In this case the customer (third party) must prove beyond any reasonable doubt that the telecommunications company was guilty of gross negligence or willful misconduct so that the compensation will not be limited to the tariff.

The best way a telecommunications company can safeguard its interests against lawsuits resulting from service interruption is by incorporating these limitations of liability in the service provision contract signed by its customers. Although this might not protect the company under all circumstances, it will provide protection in certain situations.

10.4 INSURANCE

A telecommunications company can obtain insurance to protect itself from lawsuits resulting from service interruptions. Here are two policies:

- *A general liability policy*: this might cover the liability for personal injury or property damage.

- *Errors and omissions policy*: this might cover liability for financial loss in business resulting from negligence or misconduct. Such a case might arise if it can be proved beyond any reasonable doubt that the persons involved in the fault repair failed to act professionally or that the telecommunications company did not plan beforehand how to cope with such disasters properly.

10.5   INDEMNIFICATION

A telecommunications company might decide to indemnify its directors and officers against:

- Legal expenses incurred because of their position in the company

- Amounts paid in settling company lawsuits

- Any other charges concerning these service interruption lawsuits

A telecommunications company that finds it is in its best interests to indemnify its directors and officers, must obtain insurance coverage against these risks. The insurance premium will depend on the insurance company's investigation. The insurance companies will be interested to verify that the telecommunications company has certain disaster preparedness plans, such as the following:

- *Hazard analysis plans*: the insurance companies want to insure that the telecommunications company has identified those types of problem that could cause service interruption.

- *Vulnerability analysis*: a plan to identify the vulnerabilities in the network and how to cope with them.

- *Disaster management plans*: a plan to deal with disasters if and when they occur. Among other things, an insurance company would be looking for:
  — An alternative service provider contracted to take over during a disaster
  — A standby power generator unit in case of power interruption
  — Alternative computer facilities in case of fire or power breakdown

- *Service provision contract*: the insurance companies will be interested to review the contents of the service provision contract signed by the telecommunications company's customers, to examine the liability limitations and agree to them or to amend them as they require.

When the insurance companies are satisfied that these items are available and suit their requirements, their charges for the insurance will be reasonable. Thus it is in the best interests of the telecommunications company to have a well-established disaster management plan which minimizes any losses due to service interruption.

SUMMARY

When the telecommunications organization makes an agreement with a customer that uses telephone lines to transmit information, the agreement usually stipulates what should be done in case of a disaster, i.e. when the telecommunications organization cannot provide the service for a long time or a short time. This kind of agreement usually addresses the following points:

- The organization's duty to insure continued service to its customers.
- The limits of this duty, how and when it can be held liable to the losses that may be suffered by its clients due to the interruption of the service.
- The duty of the O&M staff to insure continued service.
- The limits of this duty, how and when the managing director or even the O&M staff can be held liable for the losses that may be suffered by the organization's clients due to the service interruption.
- The formula that will be used to calculate any compensation.

In government-owned telecommunications organizations the customer cannot sue the telecommunications service provider because the government cannot be sued unless it agrees to be sued. If the government agrees to be sued, it will indicate clearly the extent to which it has consented to be sued. In the case of privately owned telecommunications companies, the customer can sue the company for losses incurred due to the interruption of the service, whether for a long time or a short time.

Telecommunications companies limit their liability by introducing certain clauses in the service agreement contract with their customers, which in effect prevents the customer from claiming the full amount of compensation due to the interruption of service, and limits the amount to be paid by the company to the tariff the customer would have paid for the duration of the interruption. Furthermore, telecommunications companies insure and indemnify their directors and O&M staff against any negligence or professional misconduct.

Before issuing an insurance policy, insurance companies will be interested to verify that the telecommunications company has created and put into force certain disaster preparedness plans, such as:

- Hazard analysis plans

- Vulnerability analysis
- Disaster management plans

Insurance companies will also be interested to review the contents of the service provision contract signed by the telecommunications company's customers, to examine the liability limitations and agree to them or to amend them as they require.

REVIEW QUESTIONS

1. Define the following legal terms: liability, insurance, indemnification.

2. What is the liability of a private telecommunications company, regarding its obligation to provide continuous service to its customers?

3. What is the liability of a public telecommunications entity, regarding its obligation to provide continuous service to its customers?

4. What are the limits of liability of a telecommunications service provider?

5. Why would it be in the best interests of a telecommunications company to prepare a comprehensive disaster management plan?

6. On what basis would an insurance company estimate the amount to be paid by a telecommunications company, to insure its staff against service interruption lawsuits?

11

Case Studies

OBJECTIVES

- Answer questions at the end of the chapter.
- Discuss how you can analyze a case study.
- Give examples of how disasters are dealt with at your workplace.
- Identify the positive and negative aspects of how disasters are dealt with at your workplace.
- Present examples of similar disasters at your workplace and be able to properly analyze them.
- Describe the nature of some of the worst disasters at your workplace.
- Demonstrate how you can reach the core of the problem and how you can make it less likely.

11.1 HOW TO DEAL WITH CASE STUDIES

Case studies offer training on how to deal with problems in a rational way before they develop into disasters. They also provide opportunities to analyze and discuss any previous disasters that may have been encountered.

11.1.1 Step-by-step analysis

Identify the problem

We identify the problem and define it precisely; we have to read the case study properly and note all the relevant points.

Study the environment

We study the problem's environment; we look at it carefully and discuss how each environmental element affects the problem.

Use analysis forms

We use a variety of forms to summarize the main points; this helps us to concentrate on the major causes and effects, making it easier to reach a sound decision. Some of the steps are listed in Table 11.1.

Differentiate symptoms and causes

We must tackle the causes of the problem, not the symptoms. A child infected with a virus may have a high temperature. A cold compress may lower their temperature and make them feel more comfortable but it probably won't kill the virus. High temperature is a symptom of the child's problem and the virus is the cause.

Evaluate the solutions

We write down some possible solutions, indicating their advantages and disadvantages in the short term and the long term, and from several viewpoints: technical, financial, administrative.

Select the appropriate solution

From our shortlist we select the solution that solves the problem most effectively according to our requirements.

Table 11.1 Problem analysis form

Item	Description
Summary of the problem	Initial conditions Triggering event Other developments:
Symptoms	1 2 3

Table 11.1 (Continued)

Item	Description
Probable causes	1 2 3
Actual causes	1 2 3
Suggested solutions (including advantages and disadvantages)	1 2 3
Cost of each solution (C_s)	1 2 3
Cost of damage (C_d)	

(continued overleaf)

Table 11.1 *(Continued)*

Item	Description			
Rationale of the selected solution	1 2 3			
Changes required in the system	1 2			
Changes required in the process	1 2			
Other changes required	1 2			
Estimated recovery percentage (R_p)				
Adjusted cost of damage, $C_{ad} = C_d(1 - R_p)$				
Merit ratio, $R_m = C_s/C_{ad}$				
Constraints	1 2 3			
Priority of the problem	very high high	medium	low very low	

Determine how to implement it

We carry out a study to determine the resources we require to implement the solution we have chosen. We use the form in Table 11.2.

Draft the report

We need to produce a draft because it is usually hard to write a well-structured report in one step. We write the report headings then fill in our comments and observations. Here are some suggestions for the headings:

- *Summary*: we summarize the main points to explain how we built up our judgment, but we do not go into details about the problem itself — that's the next stage.

- *The real problem*: we formulate the problem we have studied; a well-formulated problem is halfway to solution. We can use several forms to identify the problem.

- *Problem analysis*: we arrive at the real causes for the symptoms we have observed. We must clearly state any assumptions we have made.

- *Suggested solutions or decisions*: we look for possible solutions to the problem we have analyzed. Then we compare them to arrive at the best option, taking into consideration the technical, financial and administrative aspects.

- *Recommended actions*: we recommend the actions to be taken to implement our chosen solution. If we have identified several problems, then we might need several solutions; we need to identify which solution solves which problem.

Amend the draft report

We read and review the draft report as a whole, then make amendments to insure it is readable and understandable. By 'readable' we mean that the presentation is logical and rational so the reader will not give up reading it (Table 11.3). By 'understandable' we mean that our explanations are as simple and concise as possible and we have provided all the information required to develop an argument.

Our report must be long enough to cover the major points thoroughly, but short enough to highlight the important problems succinctly. Readers might skip sections of a long report or give up altogether (Table 11.3). We should look at the spelling and grammar and make appropriate corrections.

Word-process the report

We should word-process the report on a computer. Handwritten reports are sometimes hard to read and they usually look unprofessional. Type-written reports may also give the wrong signals to readers.

Table 11.2 Solution implementation form

Item	Description		
Summary of the problem			
Priority	very high high	medium	low very low
Merit ratio			
Suggested solution			
Decision taken			
Implementation procedure	1 2 3 4 5		
Resources needed	1 2 3		

Table 11.2 (*Continued*)

Item	Description
Implementation team members	1 2 3
Implementation team responsibilities	1 2 3
Implementation timetable	
Risks involved	1 2 3

Table 11.3 Report sections of interest to managers and the percentage that actually read them

Report section	Managers who read it
Executive summary	100%
Introduction	65%
Main text	22%
Recommendations	55%
Annexes	15%

Review the report

We read the word-processed document before submitting it. We check that it adequately sets out the problem and clearly explains our ideas on how to solve it. Even at this late stage, we might discover a point that needs clarification, a paragraph to insert or a sentence to remove. Reviewing the report is very important.

11.2 TWO SOLVED CASE STUDIES

11.2.1 Clocks and radios

A watch and clock manufacturing company experienced a drop in its revenues. The management thought this was due to the general economic situation in the country; yet when the economy picked up, the revenues were still dropping. The marketing department conducted a survey, which showed that customers preferred to buy a radio alarm clock unit instead of the company's clock products. Although the price of the manufactured clock was cheaper than the clock used in the radio alarm unit, the customers preferred the radio alarm clock since its price was lower than the cost of a separate radio and the company's clock.

The marketing department suggested that the engineering department should begin to manufacture the radio alarm clock unit at a lower price so the company could compete in the market. The engineering department thought this an overdue suggestion; had it been presented earlier then perhaps R&D, planning and manufacturing could have been executed before the company's market share had dropped. The marketing department indicated that they couldn't have predicted the situation since the computer printouts of sales were available only for actual sales, with no projections for the future.

The engineering department cautioned that entering the electronics field to manufacture the radio unit would need a lot of investment in R&D, equipment purchase, recruitment and training. The finance department indicated that the company's financial situation was still okay and that they could secure the needed loans from the banks once there was a feasibility study with which to approach them.

The problem analysis is given in Table 11.4 and the problem solution in Table 11.5.

11.2.2 Telaria TV

Background information

- Telaria is a developing country, its population is about 2 million, 40% of them in the capital. It has one TV station and one FM radio

Table 11.4 Problem analysis for clocks and radios

Item	Description

Summary of the problem	**Initial conditions**
	  Production is efficient
	  Finances are okay
	Triggering events
	  Clients buy from competitors
	  Drop in sales
	Other developments
	  Company product cannot be marketed
	  Competitor products more advanced
	  No market research available
	  Company objectives are not clear

Symptoms	Losing market share in clocks Department heads don't talk with one another Each department waits for an approach No teamwork
Probable causes	Fierce competition Obsolete product Shortsighted management Unclear objectives
Actual causes	Failing to predict new markets Producing obsolete product
Suggested solutions (including advantages and disadvantages)	Develop planning activities Produce a better product Change the line of production
Cost of each solution (C_s)	Planning Product Production line
Cost of damage (C_d)	Total loss of activity

Rationale of the selcted solution	Develop planning activities
	  Long-term solution
	  Solves problem forever
	  Incorporates other solutions

Changes required in the system Changes required in the process Other changes required	*This section is completed* *after writing the report*
Estimated recovery percentage (R_p)	
Adjusted cost of damage $C_{ad} = C_d(1 - R_p)$	
Merit ratio, $R_m = C_s/C_{ad}$	
Constraints	Availability of funds and electronics expertise
Priority of the problem	(very high) medium low high very low

Table 11.5 Solution implementation for clocks and radios

Item	Description
Summary of the problem	
Priority	very high medium low high very low
Merit ratio	
Suggested solution	
Decision taken	Develop planning activities in the company
Implementation procedure	Recruit a planning skills development expert Conduct a market research survey to identify long-term client needs Develop R&D activities Prepare and regularly update a long-term plan Set up work performance standards for each department Develop managers' planning skills Develop computer-assisted activities to help in decision making
Resources needed	External expertise to develop planning skills Training aids for the planning skills development program Time for managers to attend the program Funds for market research Funds for R&D in the engineering department
Implementation team members	Planning skills development expert Department heads Computer/IT unit head
Implementation team responsibilities	Analyze how to gain the market share within a specified time Document the production capabilities needed to regain the market share Prepare a long-term plan to meet product and client preferences
Implementation timetable	3 months
Risks involved	Department heads may not want IT Costs may be high

Note: In the Priority cell, "very high" is circled.

broadcasting station in the capital to serve the population. There are about 50 000 TV sets and about 150 000 FM receivers in the country.

- The TV transmitter lies in the valley of a medium-height hill (about 100 m high). On the top of the hill is a 50 m high steel structure carrying the antennas. There are four antenna panels arranged so that each one occupies one side of the tower, a steel structure with a

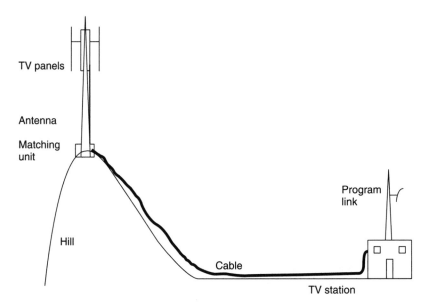

Figure 11.1 The TV station site

square cross section, so the resulting radiated field is omnidirectional to cover the whole city and its outskirts.

- A coaxial cable connects the transmitter in the valley to the antennas on the hilltop. The station layout is shown in Figure 11.1.

Problem faced

- One day the viewers were enjoying the TV program when suddenly they lost the sound and the pictures. They complained by phone but received little or no response.

- The chief engineer was watching on his office TV set, so he thought the complaints were unjustified.

Action taken

- When the level of complaints became too high, the TV station chief engineer asked the shift engineer to insure the station was running normally.

- The shift engineer measured the TV radiated power on the dummy load and found it okay. He then measured the antenna-side voltage standing wave ratio (VSWR) at the transmitter output; it was 1.2 when it should have been 1.1 or less.

- The chief engineer thought the difference in the VSWR reading was insignificant since it amounted to a difference of about 9%.

- When the chief engineer was contacted by the responsible minister, who complained that he could not see the program, the chief engineer was alarmed and asked one of the antenna technicians to climb the TV tower and look for any irregularities.

- The technician came back and told the chief engineer that all was well and asked for a prize for his good work! When asked, he declined to say what exactly he had done to deserve a prize.

- Reports came in that the northern and southern districts of the city had good reception, the eastern part had exceptionally good reception, but the western part had very bad reception. The problem here was that the minister's home was in the western district of the city, and he was furious when for the second time he contacted the chief engineer to complain.

- The chief engineer asked the shift engineer to do something. So the shift engineer realized that the pattern must have changed from omni to cardioid, which explains why parts of the city had good reception while one part had very bad reception. Thus he went up the tower and repaired the pattern, by correctly fixing the connections to the matching unit of the four panels and their distribution network.

Discussion

- Examine the various problems that faced the TV station chief engineer, the shift engineer and the viewers.

- What were the causes of the disaster, who were the people responsible and why were they responsible?

- If you were to perform the faultfinding or disaster recovery procedure, how would you go about it?

- What safety measures were neglected?

- What safety measures do you recommend so that no such disaster can happen?

- What administrative action should be taken to help those involved in the disaster?

Analysis

- The chief engineer ignored the early warning signs that there was a fault in the station, on the grounds that if he could watch the program on his office TV, the complaints must be unjustified. He was wrong because he failed to realize that he was receiving a signal from the RF output of the station not from the antenna.

- The chief engineer failed to interpret the measurements correctly, thinking that a difference of 9% was negligible. He was wrong because the voltage standing wave ratio (VSWR) was high but it was attenuated due to the long length of the cable. If he had asked the shift engineer to measure the VSWR at the antenna terminal, he would have noticed a very alarming value.

- The chief engineer bypassed the shift engineer and contacted the technician, which is administratively and technically wrong.

- The chief engineer didn't give proper instructions to the technician, he only told him to see what was wrong. Thus the real message the technician received from the chief engineer was simple: I do not know what is the problem; you go and investigate. That is why the technician asked for a prize, since he'd achieved something that the chief engineer hadn't.

- When the complaints continued, the chief engineer was forced to ask the shift engineer to intervene.

- The shift engineer thought logically and found the fault caused by the technician. A high wind had loosened the main connection of the antenna cable to the matching unit; had this connection been tightened, the problem would have been solved. Yet the technician, not knowing what to do precisely, unscrewed the four panel connections and didn't put them back in their proper order. This created a phase shift error.

11.3 SOME MORE CASE STUDIES

11.3.1 Power interruption at Telaria TV

Background information

The TV and broadcasting headquarters building is situated in downtown Telaria; it is composed of 28 stories and contains offices plus several TV and sound broadcasting studios. On the roof are situated microwave program links that carry the programs to the TV and radio broadcasting transmitting stations outside Telaria. The building is circular in plan with offices in the periphery so that the studios, which are in the middle, are shielded from the effect of outside noise and interference. The building works 24 hours all year round, with three shifts per day working all the time.

Problem faced

- One day at 0630 a fire broke out on the 26th, 27th and 28th floors of the TV and broadcasting headquarters in downtown Telaria.

- The building contains administrative offices, radio broadcasting and TV studios. Microwave program links on the roof transmit the programs to transmitting stations on the outskirts of Telaria.

- The fire lasted for six hours, although the fire brigade was immediately called to put out the fire. Since fire erupted in the upper floors, which were hard to reach by the fire brigade, the fire affected the microwave program links on the roof.

- The main problems were:
  - To restore TV and broadcasting services in time.
  - To rectify the fire damage to the power cables.
  - To resume operation as soon as possible by finding an alternative for the microwave links.
  - To rectify all other issues that resulted from the disaster.

Action taken

- The power supply was cut from the building to enable the use of water jets for firefighting.

- The civilian and military fire brigades were contacted to collaborate in putting the fire out. A helicopter unit was tasked to help in putting the fire out by spraying chemicals from above the building, to reach the higher floors.

- Three committees were initiated to look into the technical, administrative, financial and criminal aspects.

- The criminal investigation committee reported these findings:
  - There is no indication that an act of arson is involved.
  - The reason for the fire was expected to be due to leaving a butane gas cylinder open and mistakenly throwing a lighted match in the dustbin; both acts were presumably committed by a cleaning worker on the 27th floor, who died during the fire.

- The administrative and financial committee reported these findings:
  - The total cost of damage due to the fire is estimated at 16 million Telarian pounds (US$ 5 million).
  - Several fires happened in the building before that one, the most recent was one year earlier. After each fire incident a report was issued indicating the precautions to be taken to prevent another disaster from happening, yet all these precautions were ignored. (This is why several other fires happened, almost one per year.)
  - There is very little firefighting equipment in the whole building, and several floors do not have any equipment at all.
  - There is no firefighting strategy and no fire drills are conducted.

- The technical committee reported these findings:

— Negligence, and the lack of firefighting and control facilities, was the main reason behind the long time taken to put out the fire.

— Fire spread quickly from the 27th floor (where a cleaning worker left the butane gas cylinder open and threw the match in the dustbin while it was still lit) to the upper and lower floors, due to the thick carpeting used on these floors and due to the blocking of some corridors with trash paper. This helped the fire to spread more quickly, generating smoke that caused several fatalities due to choking.

— The microwave equipment installed on the 28th floor was damaged by fire, and the antennas on the roof were also damaged by the intense heat generated from below.

— Fire spread through the waste chute to the 15th and the 2d floors, whose chute outlets were stuck by wastepaper.

— The main power supply cables were improperly bracketed on the walls near the waste chute, so the intense heat affected most of them. Yet there was no indication that an overload might have caused the cables to burn. Moreover, the main power supply switchboard was intact.

• The engineering department, in collaboration with the programs department, initiated a temporary studio at the site of the TV transmitting station on the outskirts of Telaria, where the TV presenters began the programs and used some videotapes. This activity began at 1430, in time for the midday news bulletin.

• After putting out the fire, an investigation of the power cables revealed that some of them were undamaged by the fire, so they were used to operate some of the studios.

• Firefighting equipment was purchased and installed at strategic locations on each floor in the building. A fire contingency plan was drawn up to prevent similar disasters.

• A plan was formulated to install new microwave links, instead of those damaged by the fire, and orders were placed for the equipment.

Discussion

• Examine the various problems that faced the TV and broadcasting management.

• What were the causes of the disaster, who were the people responsible and why were they responsible?

• If you were to perform the disaster recovery procedure, how would you go about it?

• What safety measures were neglected?

- What safety measures do you recommend so that no such accident can happen?

- What administrative action should be taken to help those involved in the disaster?

Analysis

- The disaster involved several departments and was caused by cumulative negligence and lack of coordinated work by all those who took part in the firefighting and control process.

- The engineering department should have insured minimum interruption of the programs, since the TV morning show was scheduled to begin at 1000, so the department should have used some of the TV outside broadcasting (OB) vans to transmit the TV and radio programs to the transmitting stations on the outskirts of Telaria, instead of waiting until 1430, when a studio was established at the transmitting station's site.

11.3.2 Power overload in a megawatt transmitter

Background information

A high-power medium-wave broadcasting station transmits political programs to neighboring countries. This explains the high power used and also the criticality of any fault in the operation that may interrupt the service, which in this case will be interpreted badly on the part of the responsible O&M engineer (an interruption of 5 minutes or more means high-level investigation). To gain the required high power, the station is composed of two 300 kW transmitters connected in parallel through a high-power paralleling unit, and an antenna system that gives a gain of 2. This setup provides a total radiated power in the forward direction of 1 MW. The station is connected to a 206 m high vertical antenna system composed of a radiator insulated from ground, and a grounded director. They are properly spaced to give a cardioid-shaped radiation pattern. The station is connected to the antenna system through a 1 km long open-wire coaxial transmission line (Figure 11.2).

Problem faced

At 2100 the O&M engineer noticed that while the two transmitters, 1 and 2, were working at full power, suddenly an overload happened, the two transmitters went off air and the program broadcast was interrupted.

Action taken

- The O&M engineer tried to set the power back by operating the first transmitter, and it worked okay — all controls and meter readings

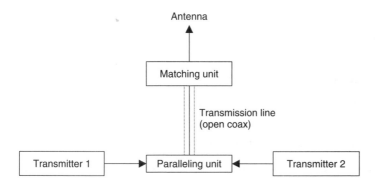

Figure 11.2 Transmitting stations: general layout

checked were okay. When energizing the second transmitter, the two transmitters failed again. Reversing the order of energizing gave the same result. Operating each of the two transmitters on the dummy load gave good results, yet it was not possible to operate the two (at the same time) on dummy loads with the paralleling unit, due to the high power involved.

- The engineer thought the problem was in the paralleling unit; he also thought that since half the total power was handled okay then there must be something in the path that was properly handling half the power but which became short-circuited at full power. Careful inspection revealed that the paralleling unit was okay.

- The engineer began to look for a fault in the open-wire 1 km coaxial transmission line. Since it was dark at night, the search took some time but the transmission line was found to be okay.

- The last part was the antenna feed itself, which was also found to be okay.

- From the beginning, the O&M engineer didn't bother to look at the director antenna's base spark gap because he thought that if the gap clearance was enough to pass half the power then there must be nothing wrong with it at full power. Yet as a final try he checked the spark gap and found a burnt pigeon (burnt by the RF voltage spark) sitting on the spark ball in the gap, reducing the effective gap width (since the pigeon at that moment was like a piece of coal).

- The problem here is that the O&M engineer had the wrong idea about the effective radiated power when one station only was on and with the paralleling unit in place; this was because, as he calculated later on, the power was only 25% of the total power, since a dummy load was automatically inserted in the circuit when one station was off, to keep the balance of the paralleling unit (Appendix E).

Discussion

- Examine the various problems that faced the engineer.

- What were the causes of the accident, who were the people responsible and why were they responsible?

- If you were to perform the faultfinding procedure, how would you go about it?

- What safety measures were neglected?

- What safety measures do you recommend so that no such accident can happen?

- What administrative action should be taken to help those involved in the accident?

Analysis

If the O&M engineer had had the correct information about the paralleling unit, which he should have obtained as part of a crisis prevention scenario, he would have suspected that the very low emitted power couldn't cause the short circuit, and then he would have suspected a problem at the antenna base in the first place, reducing the time taken to handle the problem. The failure problem developed into a disaster because the time taken to clear the fault was far more than the 5 minutes allowed, and the O&M engineer was investigated.

11.3.3 Early morning problems at a megawatt transmitter

Background information

A high-power (megawatt) medium-wave broadcasting center contains two broadcasting stations, A and B, that transmit two different political programs to neighboring countries, which explains the high power used and the criticality of any fault in the operation. A fault may result in service interruption, which in this case will be badly interpreted on the part of the responsible O&M engineer (an interruption of 5 minutes or more means high-level investigation). To gain the required high power, each broadcasting station is composed of two 300 kW transmitters, 1 and 2, connected in parallel through a high-power paralleling unit, and an antenna system that gives a gain of 2. This setup provides a total radiated power of 1 MW in the forward direction for each station. The output stage of each transmitter is composed of three forced air-cooled output tubes, 100 kW each (1.5 m high, 15 kg weight), connected in parallel (Figures 11.3 and 11.4).

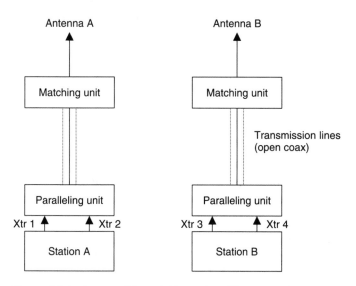

Figure 11.3 Transmitting stations: simplified circuit diagram

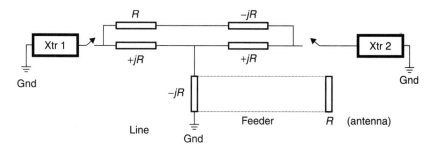

Figure 11.4 Paralleling unit: simplified circuit diagram

Problem faced

One day at 0500, when the transmitter technician was preparing the station to be operated for the early morning transmission, he noticed that transmitter 2 couldn't be operated. He performed the usual routine tests but couldn't find the fault, so he alerted the O&M shift engineer. The program was due to begin at 0600, so there was plenty of time, at least that's what the technician thought.

Action taken

- The O&M shift engineer began by checking transmitter 1 to ascertain that it was working properly, and he was pleased to find that it delivered maximum power. He knew that the transmitted power would be 50% if the first transmitter worked on the antenna alone.

But if it worked alone with the paralleling unit in circuit, the power would be 25%. Thus he chose the first option, 50% power, thinking that the fault would be cleared quickly enough for him to bring transmitter 2 into service in time, knowing that in this case he would be obliged to shut down transmitter 1 for a few minutes to insert the paralleling unit before putting the two transmitters on again.

- He then began checking transmitter 2 and realized that the problem was due to a faulty 100 kW output tube.

- He asked the technician to get a new tube from the store and replace the faulty one. This took some 15 minutes to be completed, since the technician shut off power to transmitter 1, made the change then put the power back on.

- Transmitter 2 was operated but could not attain full power. The shift engineer checked the meter readings again and came to the conclusion that the new tube was probably faulty too. Yet before asking for another tube, he rechecked the circuit one more time and became confident that he needed a new tube.

- Knowing that there was a problem with stores credibility (since the tube he obtained was in fact faulty), and fearing to lose any more time, he decided to ask the technician to get him one of the tubes already in use at broadcasting station B, since its program was due to begin at 0700.

- He asked the technician to operate one transmitter from station B and make sure that all was well. After that he shut it down and got himself one of its final-stage tubes. In this case, he thought, there will be no doubt that the tube is okay.

- The shift engineer suddenly realized that time was quickly disappearing and that only 8 minutes were left before the program would begin.

- To save time, the shift engineer made a change and operated transmitter 1 on the paralleling unit so that the preprogram tone could be heard by anyone monitoring the station at headquarters, and also so that transmitter 2, once repaired, could be inserted in circuit without interruption of service. He thought he would get the faulty tube out of circuit while waiting for the new one.

- The shift engineer operated the mechanical interlock handle of transmitter 2 to disconnect it from the paralleling unit, knowing that the interlock disconnects the paralleling unit and clears the transmitter doors mechanically, so if the interlock is not fully disconnected the doors will not open. Yet he did not check to hear the click of the mechanical interlock. He then opened the door of the final stage to reach the faulty tube, knowing that there is an electric control interlock in the upper corner of the door that interrupts all voltages when the door is opened.

- The door to the final-stage tubes opened smoothly, and the shift engineer reached with his right hand to the tube casing (anode) to open the casing lock, so he could change the tube. Then to his amazement, he noticed that sparks extended from his fingers to the tube casing.

- He tried to pull away his hand but it was paralyzed; his right side was stuck to the tube casing and so was his leg. He felt sure that he would die. He then thought of interrupting the current path by jumping upwards and kicking himself backwards to the floor, away from the tube. He succeeded and was saved, although he suffered burns to his hands, right side and leg.

- The new tube was connected after switching off the first transmitter for a few minutes, and the station went on the air at full power just one minute before the program was scheduled to begin.

- The failure problem developed into a disaster because the O&M engineer was seriously wounded and could have been killed, although the transmitter was put on air before the time of the program. Yet the monitoring team at headquarters noticed the low emitted power and the interruption before beginning the program, thus the O&M engineer was also investigated. The catch here is that not only the O&M engineer made mistakes, but also the maintenance team, and the station manager made several mistakes; all of them led to the disaster.

- Investigations revealed the following:
  — The night before the disaster, the maintenance team was working in this station. To speed up the work, they disconnected the mechanical and electrical interlocks and forgot to put them back after they had finished.
  — The first tube taken from the substore was actually faulty.
  — Connecting the transmitters in parallel without disabling the mechanical interlock gives a high-voltage high-frequency path from the RF output tubes of each transmitter to the other, since they all feed into a unified tuned circuit in this case.

Discussion

- Examine the various problems that faced the shift engineer.
- What were the causes of the accident, who were the people responsible and why were they responsible?

- If you were to perform the faultfinding procedure, how would you go about it?

- What safety measures were neglected?

- What safety measures do you recommend so that no such accident can happen?

- What administrative action should be taken to help those involved in the accident?

Analysis

The O&M engineer

- He began his work without reading the station logbook. Had he read it, he would have realized that a maintenance team was working on the station from 0000 till 0400 that day. He should have contacted them to make sure the station was in a working condition after maintenance, and to check for any irregularities they might have noticed.

- He opened the mechanical interlock handle and failed to notice there was no audible sound of circuit interruption. This means that the mechanical interlock was disabled. Furthermore, he failed to notice that the electrical interlock in the door was disconnected; he could have noticed its disconnection if he had used the earthing stick before touching anything. But he didn't use the earthing stick because he thought he was not going to touch any capacitors (which store electrical energy), so there was no need for the stick—improper reasoning and deduction.

- He reached for the tube casing with his right hand; the rules say that if you are a right-handed person you should reach with your left hand, and vice versa.

- He was anxious to operate the station well before the program's starting time, so he made a mistake by connecting the working transmitter to the paralleling unit, enabling RF high voltage to flow from the working transmitter 1 to the faulty transmitter 2, where he was working.

The maintenance team

- They disabled the mechanical and electrical interlocks on the station (an improper action) and failed to put them back in order (another fatal mistake) before they left the station.

- They performed the maintenance and found a fault that they couldn't solve, so they left the station and went home (a third fatal mistake).

Although the regulations state that they have to insure the station is in full operation before they can leave, they just ignored the rules.

The station manager

- He didn't control the inventory of the main component store, which enabled a faulty tube to be put in the store and to be mistakenly used as a new one, giving the O&M engineer a very hard time in his faultfinding endeavor.

- He didn't control the maintenance operations properly, which allowed the maintenance team to commit several fatal mistakes.

- After the accident happened, he wouldn't initiate legal action against the maintenance team, although a shift engineer could have lost his life due to negligence on the part of the maintenance team.

The O&M engineer should have

- Read the logbook, and if he had any doubts, he should have contacted the maintenance team to clarify the situation. Reading a few lines would have taken very little time and may have saved him from many complications.

- Known that maintenance was going on the night before; he should have been extra cautious while performing his duties.

- Worked safely by applying safety rules and by using the earthing stick.

- Inspected the contents of the main store at regular intervals, or urged the station manager to carry out inspections.

- Checked the work of the maintenance team on future occasions to insure the incident was a one-off not a regular occurrence with this maintenance team. If everything seems okay, it helps to restore confidence.

11.3.4   Unscheduled detonation at a coal mine

Background information

A coal mine has been in operation for a long time; the underground digging has been going on in several directions according to the anticipated thickness of the coal layer. To speed up the process, controlled explosions have been utilized. Formerly organized by a team of explosives experts, the mine management has now decided that its own staff have gained enough experience to do the job themselves.

Problem faced

The new team distributed the explosives in their grooves within the mine walls, then connected the detonator with a long cable to the pulse-generating unit that gives the signal to operate the detonator. To insure utmost safety, the team leader decided to make the explosion next morning, which was a public holiday, so nobody would be in the mine. But that evening the detonator was operated prematurely and for no clear reason. The explosion killed several mine workers.

Action taken

- A team of experts was tasked to investigate the reason for the disaster.

- The experts found that the mine explosives team connected the explosives with the detonator and the pulse-generating unit properly. They further found that on the night of the explosion there were several thunder strikes in the area. Maybe, they thought, a pulse was generated by the thunder in the connecting cable and went through the cable to the detonator, causing the disaster. But the pulse-generating unit was located in the middle of the mine chute, and this assumption was unlikely.

- Further investigation revealed that the cable connecting the detonator to the pulse-generating unit was an unshielded cable and it passed near the main electrical power supply cable feeding the mine.

- The thunder strikes generated pulses in the main electrical supply cable, and by conduction the pulses were induced in the detonator connecting cable, causing the explosives to go off and creating a disaster.

Discussion

- Examine the various problems that faced the explosives team and the mine management.

- What were the causes of the accident, who were the people responsible and why were they responsible?

- If you were to perform the faultfinding procedure, how would you go about it?

- What safety measures were neglected?

- What safety measures do you recommend so that no such accident can happen?

- What administrative action should be taken to help those involved in the accident?

Analysis

- The mine management should have kept the explosives experts with the mine team for an appropriate period, after which the mine team should have applied for a license to operate as an explosives team. Only when they were granted the license could they be given the responsibility for operating with explosives.

- Shielded cables should always be used for any pulse-carrying cable, so that nothing can be induced in the cable wires from an external source of electromagnetic interference.

- Environmental conditions should be taken into consideration, especially in a country like Zimbabwe, where the incidence of lightning strikes is about 90 days per year and the recorded lightning fatalities amount to 300–350 persons per year. The effects of lightning strikes should be considered very seriously.

11.3.5 Power cable explosion in a mining tunnel

Background information

A coal mine has been in operation for a long time; the underground digging has been going on in several directions according to the anticipated thickness of the coal layer. To speed up the process, more drilling machines are needed, hence more electrical power. The distances covered are long and the cables are stretched for long distances. Since the power consumption of the drills is large, and it is a long distance from the generating station to the drills, the cable diameter has to be large to reduce the voltage drop and the heat generated.

Problem faced

- One day the mine management was shocked to learn that a large explosion had occurred in the mine, killing several mine workers. Preliminary investigations indicated that one of the power cables was the cause of the explosion, the fire and the hydrogen chloride gas that accompanied it.

- The mine engineering department confirmed that the cable was designed according to the IEE Wiring Regulations (16th edition) and that an overload was very unlikely to have caused the disaster.

Action taken

- A team of experts was tasked to investigate the reason for the disaster.

- The experts found that:
  — The cable was running in a long tunnel with very poor ventilation, and the environmental temperature was higher than recommended.

— The percentage of coal powder present in the mine's air was more than the recommended figure.

— The cable was designed according to the Wiring Regulations but didn't take into account the effect of the high temperature in the tunnel and the large concentration of coal powder in the mine's air.

- The experts concluded that the cause of the explosion was the precipitation of coal onto the cable. This contributed to reducing the heat emission from the cable, and it was compounded by the high ambient temperature of the tunnel.

- Further investigation revealed that the cable length was increased by the operations department, without the knowledge of the engineering department, to reach a length about 25% greater than the maximum calculated by the engineering department.

Discussion

- Examine the various problems that faced the mine management.

- What were the causes of the accident, who were the people responsible and why were they responsible?

- If you were to perform the faultfinding procedure, how would you go about it?

- What safety measures were neglected?

- What safety measures do you recommend so that no such accident can happen?

- What administrative action should be taken to help those involved in the accident?

Analysis

- The mine management should regularly have monitored the temperature recordings at different areas in the mine so that the temperature used for calculating the cable rating would hold while working with the cable.

- The mine management should regularly have monitored the coal powder concentration recordings at different areas in the mine so that the rate of precipitation used for calculating the cable rating would hold while working with the cable.

- The mine management should have issued strict rules to forbid extending the length of the power cable more than the calculated length without the authorization of the engineering department.

- The engineering department should regularly have monitored the use of the cable to ensure it was properly utilized. They should also have checked the precipitation of coal powder on the cable to give early warning that something disastrous might happen if things were not returned to normal.

- The operations department should have consulted the engineering department before messing with the power cable.

11.3.6 Earth station upgrade with ministerial intervention

Background information

Situated in one of the developing countries, there was an Intelsat standard A station built in 1975 and operating with Indian Ocean satellites. When Intelsat announced a changeover from type 4 to type 4A, the earth station had to be upgraded.

Problem faced

- The station engineer discussed the requirements with the director of telecommunications and they agreed to order the relevant items from the manufacturer.

- The operation and maintenance of the earth station was given to a foreign company, and the ministry was getting 25% of the revenue in return. The company was not consulted in the matter of the equipment upgrade since it was outside their contractual obligations.

Action taken

- The director of telecommunications submitted a request to the minister of communications, so that funds could be allocated from the ministry of finance, since the department had no funds allocated for this purpose.

- The O&M company was not happy to order the equipment from the manufacturer without their consent, so they contacted the ministry of finance to inform them that the funds needed were more than the actual amount, implying that the director of telecommunications didn't have the needed expertise to handle the matter.

- The matter was referred to the ruler's office for guidance; this took several weeks.

- The Intelsat deadline was approaching, so the director of telecommunications tried to meet the minister of communications to ask him to speed up the matter before the deadline was reached and the station became inoperative. The minister told him to try meeting the ruler's technical consultant.

- The director of telecommunications met the ruler's consultant several times, in the presence of the O&M company's representative. It was clear from these meetings that the O&M company was trying to postpone taking any action. As a final step, the director of telecommunications sent Intelsat an urgent fax requesting them to consider postponing the changeover since the country was not yet ready.

- On the day of the changeover, the director of telecommunications was summoned to the minister's office, where he found the director of the O&M company and his interpreter.

- The O&M company manager told the minister that the disaster was due to the incompetence of the director of telecommunications. He reminded the minister that his company had raised this point several times before but had received no action from the ministry, and this in his opinion was the cause of the disaster.

- The director of telecommunications was asked by the minister why the funds were not allocated. He replied that he had asked all concerned for the funds and stressed that the minister himself was requested to take action several times. The minister replied, 'But you never indicated that it was urgent or dangerous.'

- The O&M company manager respectfully suggested that the ministry should terminate the services of the director of telecommunications with immediate effect, so that his company could begin to rectify the situation.

Discussion

- Examine the various problems that faced the director of telecommunications, the minister and the O&M company manager.

- What were the causes of the disaster, who were the people responsible, and why were they responsible?

- If you were to perform the earth station upgrade, how would you go about it?

- What safety measures were neglected?

- What safety measures do you recommend so that no such accident can happen?

- What administrative action should be taken to help those involved in the accident?

Analysis

- The main problem causing the disaster happened because the director of telecommunications didn't consult the O&M company before asking for the new equipment. He did so on the grounds that this was outside their contractual obligations, which was correct. Yet he could have informed the company with his intent without asking them for action, in this case he would have defused any potential resentment.

- The day before the changeover, the director of telecommunications was visited by the O&M company manager and asked what he was going to do. He replied, 'Nothing, I have done all I can without success. It is now up to God to save my neck.'

- The director of telecommunications tried all avenues to get the funds allocated without success, in fact there was nothing more he could do.

- On the day of the changeover at 0800, the director of telecommunications was informed by the ministry's station manager that a telex was received from Intelsat delaying the date of changeover for three months since many countries had asked for that. The director of telecommunications, knowing that he would be summoned to the minister's office, ordered the ministry's earth station manager to keep the telex with him and not to show it to anybody until after 1000.

- At the minister's office, when the O&M manager suggested that the director of telecommunications should lose his position for mishandling the upgrade and for rendering the station inoperative, the director of telecommunications challenged him to prove his words and accused him of misinforming the minister. He asked him to check his statement with his company's O&M staff at the earth station. At that time the telex was shown to the staff at the earth station and the O&M manager apologized for not having firsthand information. Thus the director of telecommunications was saved, in this case by an act of God.

11.3.7 Site selection for a time-critical earth station

Background information

An Intelsat standard A station was to be built in one of the developing countries to broadcast the 1975 regional sports festival, held every four years. The earth station would use Indian Ocean satellites. There was a tight schedule to build it, test it and put it into service.

Problem faced

- The station engineer, the director of telecommunications and the consultant inspected the different sites available within a diameter of 30 km from the main telecommunications center of the country and the main TV and broadcasting department.

- The supreme planning development committee was consulted, and accordingly three options were selected for the final selection by the ministry of communications and the ruler's office.

- A site belonging to the government was chosen, west of the telecommunications center at a distance of 15 km, and building began.

- The first thing to be built was the antenna foundation; this was to carry a weight of say 30 tonnes.

- When building preparations began, a man claiming that he was the landowner asked the ministry to pay compensation for his land. He was given what the ministry deemed the right sum of money.

- After two months the landowner realized that he could have charged more money for his land, had he known it was for that project. He then began to raid the site to terrorize the workers and the foreign experts working on the installation.

Action taken

- Having been informed of the situation, the minister ordered the director of telecommunications to use the site that was their second option, 35 km south of the telecommunications headquarters building.

- Building equipment and materials were transferred to the new site.

- New designs had to be prepared for the microwave link that was to carry the earth station output to the telecommunications center and to the TV and broadcasting center.

Discussion

- Examine the various problems that faced the director of telecommunications, the station manager and the consultant.

- What were the causes of the disaster, who were the people responsible and why were they responsible?

- If you were to perform the site selection procedure, how would you go about it?

- What safety measures were neglected?

- What safety measures do you recommend so that no such accident can happen?

- What administrative action should be taken to help those involved in the accident?

Analysis

- The supreme planning development committee gave the ministry inaccurate information about the ownership of the land. The director of telecommunications understood from the neighboring landowners that this site didn't belong to the government, yet he couldn't do anything since he had the assurances of the supreme committee. He should have advised the minister before he signed the contract.

- When the landowner agreed to sell the land to the government, he should have been informed by the ministry of the reason for the purchase. When the contract had been signed, it should have been enforced, but this didn't happen.

- The cost of changing the site was enormous, since the antenna foundation was 30% of the total cost of the buildings. Added to that was the cost of increasing the microwave link power, making the total around US$ 1.5 million.

- The completion date could not be put back because the Sports Festival could not be rescheduled. The delays during site selection meant that construction staff had to work overtime to meet the deadline. This increased the costs on the project.

11.3.8 The case of the injured welding technician

Background information

A high-power (1 MW) medium-wave broadcasting center contains two broadcasting stations, A and B, each transmitting different political programs to neighboring countries. This explains the high power used and the criticality of any fault in the operation. A fault may result in service interruption, which in this case will be badly interpreted on the part of the responsible O&M engineer (an interruption of 5 minutes or more means high-level investigation). To gain the required high power, the station is composed of two 300 kW transmitters connected in parallel through a high-power paralleling unit, and an antenna system that gives a gain of 2. It provides a total radiated power in the forward direction of 1 MW. The station is connected to a 206 m high vertical antenna system composed of a radiator insulated from ground, and a grounded director. The antenna elements are properly spaced to give a cardioid radiation pattern. The station is connected to the antenna system through a 1 km long open-wire coaxial transmission line.

When TV was introduced in the country, it was decided to install the TV antenna panels on top of the 206 m vertical radiator of the AM station. This meant that the installation work would take place while the AM transmitter was operational.

Problem faced

- Work began on the TV station, which was located in a room immediately under the AM vertical radiator.

- To shield the TV transmitter equipment from the high RF field of the AM transmitter, the TV transmitter room had to be magnetically shielded by three successive steel, tin and copper panels, with each shield welded together to create a Faraday shield around the room.

- The welder who was assigned the job was supposed to work while the transmitter power was on. It was decided that the distance between the welding site and the transmitter was safe enough.

- When the welder was going to weld near the antenna feed, the transmitter was switched off for a few hours so the welder could work safely. At that time, one of the transmitter technicians put the power on, and the welder was subjected to a very high RF voltage.

Action taken

- The shift engineer was informed that the welder was injured due to the high RF voltage; he immediately ordered a car to take the welder to the nearby hospital for medical treatment.

- An investigation was conducted to discover how the accident happened.

Discussion

- Examine the various problems that faced the shift engineer who was asked to conduct dangerous work while the station was on.

- What were the causes of the disaster, who were the people responsible and why were they responsible?

- If you were to perform the task of installing the TV station near the medium-wave station, how would you proceed?

- What safety measures were neglected?

- What safety measures do you recommend so that no such accident can happen?

- What administrative action should be taken to help those involved in the accident?

Analysis

- The investigation revealed there were no interlocks available to prevent the station operating when the welder was supposed to work. Furthermore, there were no regulations governing power switch-on.

- The welder was not wearing protective clothes and he didn't use the earthing stick before he began work.

- The technician who switched on the station had very bad personal relations with the welder, but there was no evidence of a crime.

- No warning signs were available at the site where the welder was working.

11.3.9 A catastrophic software flaw at AT&T

Background information

AT&T normally carries about 110 million calls per day. The worst service disruption that ever happened to AT&T was on January 15, 1990; it affected the service for almost nine hours.

Problem faced

- On Monday January 15 at 1430 hours, one of AT&T's toll switches in New York City developed a fault.

- This fault blocked about 50% of all AT&T traffic for a period of about nine hours.

- The services affected were long-distance calls, 800 numbers and software-defined services

Action taken

- The problem that occurred in the AT&T toll switch activated normal fault recovery routines within the switch, which meant the switch would be out of service for a few seconds.

- The connecting switches changed their program automatically to indicate that no new call should be directed to the New York exchange until the fault recovery process had been completed.

- A few seconds later, when the fault recovery process had been completed, the New York exchange was back in operation and signals were sent to the connecting exchanges to change their program to indicate that the New York exchange was functioning again.

- Due to a software flaw, the direct node link (DNL) processor was taken out of service. The duplicate DNL took the traffic load immediately, but due to the software flaw it was also taken out of service, isolating the switch from the signaling network.

- The unstable condition continued because of the random nature of the failures and the heavy traffic load.

- The AT&T engineers tried some standard procedures to reestablish the functionality of the signaling network, but they proved unsuccessful.

- The engineers knew they were facing a totally unfamiliar problem, so they began to monitor the pattern of error messages in search of the fundamental cause.

- AT&T discovered the cause was a software problem, so they isolated the program and stabilized the network.

Discussion

- Examine the various problems that faced AT&T.

- What were the causes of the disaster, who were the people responsible and why were they responsible?

- If you were to perform the task of faultfinding this problem, how would you go about it?

- What safety measures were neglected?

- What safety measures do you recommend so that no such disaster can happen?

- What administrative action should be taken to help those involved in the accident?

Analysis

- The software problem (software flaw) was introduced into all AT&T toll switches in the network during a routine software upgrade.

- The updated software was introduced because it would reduce the number of calls lost during the reestablishment of normal signaling between two offices after the removal of signaling management control.

- When confronted by the fault, the engineers resorted to standard solution methods. They wanted to clear the fault as soon as possible, so instead of stopping for a few moments to think about the peculiar nature of the fault, and then trying some innovative solutions, they began performing some standard procedures. These procedures turned out to be useless and they cost valuable time and money.

- The updated software was tested in the laboratory under controlled conditions. It would have been better to test it under the worst-case scenario, by assuming that all conceivable problems would happen simultaneously.

- Later on the engineers were able to reproduce the problem in the laboratory. They were able to correct it, test it and restore the backup signaling network.

11.3.10 The case of the abandoned videodisc

Background information

In the 1960s Radio Corporation of America (RCA) was under the chairmanship of David Sarnof; it was a giant electronics company with a dominant position in the US market. In the 1970s RCA began developing its videodisc to record sound and vision. The project was abandoned after 15 years of development, costing some US$ 500 million, and the sale of 500 000 units at US$ 1000 per unit. At that time the videodisc player was sold at a price of US$ 500 while the videocassette player/recorder was sold for US$ 1000. This is why RCA thought the videodisc would eventually replace the videocassette.

Problem faced

- The Japanese developed and mass-produced the magnetic recording/playing head, and secured the necessary precision.

- The price of the videocassette recorder/player fell to US$ 300.

- Sales of the videocassette recorder/player have doubled every year since 1981.

- Sales of the videodisc dropped; only 300 000 units were sold.

Action taken

- RCA reduced the price to US$ 200 but this didn't help.

- RCA had to abandon the videodisc project.

Discussion

- Examine the various problems that faced RCA.

- What were the causes of the disaster, who were the people responsible and why were they responsible?

- If you were to perform the task of developing the videodisc, how would you go about it?

- What safety measures were neglected?

- What safety measures do you recommend so that no such disaster can happen?

- What administrative action should be taken to help those involved in the accident?

Analysis

- The investigation revealed that the customer requirements or preferences were not only to see recorded material but to be able to record their favorite TV programs.

- RCA knew that recording material from TV is illegal and it assumed that would deter people, hence it decided not to develop a read/write videodisc.

- RCA's competitors were driven by customer requirements, and they were willing to enter into lengthy legal battles to get approval for manufacturing the videocassette recorder/player; it took them years but eventually they won.

- RCA lost large sums of money because of its misconception.

11.3.11 Disaster in Telaria: government communications

Background information

Telaria is a developing country, its population is about 2 million, 40% of them in the capital. It has one TV station, one FM radio broadcasting station in the capital to serve the population, and one telephone exchange. All the government ministries are situated in neighboring buildings (a complex) in a suburb of the capital. The ministries are supplied with telecommunications facilities through a 600-pair cable with plastic sheathing.

Problem faced

- One day at 0700, when they were beginning their work at the different ministries, the employees discovered that all telephone communications were extremely bad. A few moments later, all communications were interrupted.

- The general manager of the Telaria telecommunications authority was informed and he sent a team of engineers and technicians to clear the fault.

Action taken

- Using the meggar method, the team found that the cable was cut at a distance of 1 km from the government complex. From the cable layout maps, they ascertained that it was laid at a depth of 80 cm under the ground.

- They dug down to 80 cm but the cable wasn't there.

- A metal detector confirmed that the cable was there. So the engineer ordered digging to resume.

- They encountered the water table at 2.0 m deep but still no cable.

- The metal detector again confirmed there was a cable. The engineer installed a water pump to suck out the water and the digging resumed.

- A sheared cable was finally uncovered at a depth of 2.5 m.

Discussion

- Examine the various problems that faced the telecommunications company's management.

- What were the causes of the disaster, who were the people responsible and why were they responsible?

- If you were to perform the disaster recovery procedure, how would you go about it? What deductions would you make?

- What safety measures were neglected?

- What safety measures do you recommend so that no such accident can happen?

- What administrative action should be taken to help those involved in the disaster?

Analysis

- The cable was found laid inside heavy concrete ducts.

- The cable was laid 30 years ago and during this period the level of the groundwater increased due to the inefficient sewage system in the capital.

- Inspecting the cable diameter at the point of shear indicated it was 0.25% of the original diameter. This means the copper wires had become elongated before yielding and interrupting the service.

- It was also revealed that the different ministries had complained many times about the poor standard of communications but the telecommunications company ignored the complaints.

- The cable was found at a depth of 2.5 m and in a sheared condition; this revealed several points:
  - No soil survey was conducted before laying the cable in the trenches.
  - The cable was laid in concrete ducts; with wet soil the weight of the duct was very high, leading to sagging.

— When the water reached higher levels, the sagging increased due to the weight of the ducts, taking the cable down further.

— During this process the quality of service (QoS) decreased and complaints were filed, but they were ignored by the telecommunications company.

— The cable became cut when its diameter reached the yielding point, and this interrupted all communications.

• The problem would have been eliminated from the beginning if proper procedures had been followed, such as:

— Surveying the cable bed in the trench all along the cable route.

— Using plastic ducts instead of concrete ducts.

— Asking the planning authorities about the expected risks along the cable route.

— Handling the warning signs properly.

— Conducting regular QoS tests and properly analyzing the results.

— Using better quality telephone cables.

— Providing a relief supply such as a microwave link.

11.3.12   Disaster in Telaria: airport communications

Background information

Telaria is a developing country, its population is about 2 million, 40% of them in the capital. It has one TV station, one FM radio broadcasting station in the capital to serve the population, and one telephone exchange. The airport is supplied by a 50-pair telephone cable with paper wrapping and lead sheathing; this comes from the Telaria Civil Aviation Authority (TCAA). An alternative supply is secured through a microwave link (Figure 11.5a).

Problem faced

• One day at 0100 the railroad union organized a strike and ordered the stoppage of all trains leaving the capital.

• At 0500 the same day all telecommunications services were interrupted at the airport. The political leadership interpreted this as an act of solidarity with the railroad union.

• The general manager of the Telaria telecommunications authority was informed. He assured the political leaders that it looked like an isolated incident unconnected with the railroad union. He immediately dispatched a team of engineers and technicians to clear the fault, and to investigate how the two routes (main and alternative) could fail at the same time, creating such an unexpected disaster.

(a)

(b)

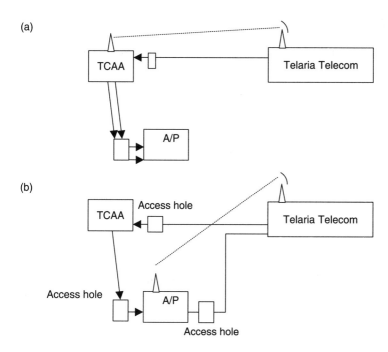

Figure 11.5 Airport communications: (a) original setup, (b) final setup

Action taken

- Using the meggar method at the TCAA building, the team found that the cable was earthed at a distance of 200 m from the airport building. From the cable layout maps they identified the nearest access hole.

- Inspecting the access hole revealed it was full of sewage water.

- A pump was requested to suck out the water and the cable was replaced with a new one, since it was full of old soldering joints.

- The alternative route (microwave link) had a terminal at the TCAA; from there the lines were connected to a 50-pair telephone cable with paper wrapping and lead sheathing that went in parallel with the main route through the water-filled access hole.

Discussion

- Examine the various problems that faced the telecommunications company's management.

- What were the causes of the disaster, who were the people responsible and why were they responsible?

- If you were to perform the disaster recovery procedure, how would you go about it? What deductions would you make?

- What safety measures were neglected?

- What safety measures do you recommend so that no such accident can happen?

- What administrative action should be taken to help those involved in the disaster?

Analysis

- The reason for terminating the microwave link at the TCAA was because the exchange was there; this was logical from the TCAA's viewpoint. After the disaster, the microwave link terminal was transferred to the airport itself.

- The very old type of cable was chosen because it was cheap. After the disaster it was changed to a jelly-filled cable — the best available option at that time.

- The condition of the cable indicated that it went faulty several times before (so many joints) and nobody reported the inherent risk of passing the two routes through the same access hole.

- For security reasons a third cable was installed directly from the telecommunications company to the airport (Figure 11.5b).

SUMMARY

When analyzing case studies, begin by identifying the problem then study the environment to determine its effects on the problem. The analysis is simplified by using a problem analysis form. Use another form to discuss the implementation strategy.

REVIEW QUESTIONS

1. When considering disasters, how might the actual causes differ from the probable causes?

2. What are the steps that should be followed when analyzing a case study?

3. What are the steps that should be taken when preparing a case study report?

4. In the clock and radio case study (Section 11.2.1) what was the real cause of the problem? And what was the probable cause? Discuss what made you decide those causes.

5. In the Telaria TV case study (Section 11.2.2) what was the real cause of the problem? And what was the probable cause? Discuss what made you decide these causes.

6. How can you use the problem analysis and solution forms to solve some of the case studies in Section 11.3?

7. How can you modify the problem analysis and solution forms to help you solve a case study where the available information is insufficient?

12

Conclusion

It is all very well to write a book.
But can you waggle your ears?

J. M. Barrie, who could, to
H. G. Wells, who couldn't

OBJECTIVES

- Answer questions at the end of the chapter.
- Discuss the main ideas of disaster management in telecommunications systems.
- Present examples of how to prepare a disaster management plan at your workplace.
- Describe the preliminary report on the disaster management plan and explain how you might present it to your management.
- Give examples of how to implement the disaster management plan at your workplace.

12.1 INTRODUCTION

What we have been striving to achieve is to prepare, test and implement a management and recovery plan that can save our organization from disaster. Figure 7.1 emphasized how we must try to prevent the disaster from occurring in the first place. But if we cannot do that, we should try to minimize its effects by applying a properly formulated plan. An effective disaster management and recovery plan is not something to assemble hastily then forget. It is a living document that must be updated, tested and exercised regularly. This chapter ties together all our work so far.

12.2 OBJECTIVES AND GOALS

12.2.1 Objectives

- To prevent disasters.
- To recover from disaster with minimum losses.
- To get back to normal as quickly as possible.

12.2.2 Goals

- Insure the safety of the employees and all those on the premises during the disaster.
- Conduct vulnerability search, analysis and rectification to identify critical areas that affect business operations of the organization (Chapters 4 and 8).
- Identify the most probable causes for a disaster in the organization (Chapter 3).
- Initiate and activate a disaster management and prevention plan (Chapter 4).
- Minimize the duration of any disruption, and resume normal operation as soon as possible (Chapter 5).
- Minimize the resulting damage and losses (Chapters 9 and 10).
- Coordinate all efforts for effective, efficient and synchronized disaster management and recovery (Chapters 6 and 7).
- Make the plan as simple as possible, so it is fully understood and properly implemented.
- Train the disaster management team on the plan regularly.

12.2.3 Management's viewpoint

- The plan should provide management with a comprehensive understanding of the grave consequences of not having a plan.
- It should define clearly the recovery requirements and resources needed to achieve its goal.

- It should contain historical data documenting the impact of service loss (long-term and short-term) on key functions.

- It should focus on disaster prevention as well as on disaster management and impact minimization.

- It should give the management several alternatives to consider, indicating how they can be integrated into the organization's ongoing planning.

- Once convinced, management should be committed to the plan and provide the necessary funds and support.

12.3 THE PLANNING PROCESS

12.3.1 Initial steps

The initial steps are used to give the disaster management team a better understanding about the existing and projected activities of the organization. Here are the relevant tasks:

- To conduct several interviews to assess the security measures and the business impact analysis.

- To develop a preliminary policy for disaster recovery.

- To explain to all participants the importance of the planning process.

12.3.2 Vulnerability search

Vulnerability search, analysis and rectification forms the basis of all disaster avoidance and recovery in the telecommunications industry. Not only does it help to protect the organization, but it also reduces the costs when a disaster does occur.

Vulnerability search takes a careful and honest look at all possible vulnerabilities in the organization. Vulnerability analysis determines the levels of disaster inherent in all the vulnerability areas then considers how to reduce them. Vulnerability rectification means taking action to reduce any risks until their levels are practically acceptable.

There are several types of vulnerabilities depending on the risk they pose to the organization. Table 6.4 details the different types along with the proposed search, analysis and rectification work.

Security assessment is an important element in the vulnerability search. If we take a computer company as an example, the security management would look into the following areas:

- Personal and operating security

- Backup and contingency planning

- Security planning and administration
- Database security
- Systems and access control software security
- Data and voice communication security

Here are some of the disasters that can be considered:

- Telecommunications network failure
- TV broadcasting station failure
- Building facilities failure (power disasters)
- Fire
- Flood
- Earthquake
- Tornado
- Hurricane
- Arson
- Security breach
- Hardware or software failure
- Financial failure
- Other disasters

12.3.3 Disaster impact analysis

Assessing the impact of the disaster on the business operations enables the disaster management team to identify critical systems, processes and functions that have the most pronounced effect on the organization if it is hit by a disaster. The economic evaluation is discussed in Chapter 8.

Consider the length of interruption the organization can tolerate during a disaster; this will determine the speed of intervention and will affect the total cost of the prevention program. This analysis will identify the critical systems, processes and functions. It will assess the resources required to support critical services. It will confirm the information gained from the vulnerability search exercise.

12.3.4 Disaster management requirements

Get a general idea of what resources are required to recover from a disaster, then use it in evaluating the different recovery strategies. These recovery strategies should be divided into immediate, short-term, medium-term and long-term. Conduct an inventory of the resources already available at the organization. This topic is discussed in Chapter 7.

12.3.5 Development of the plan

Define and document the components of the disaster recovery plan. To minimize the risks of disaster, it may be necessary to recommend some changes to the way the end user is performing their work. The necessary actions during a disaster should be clearly defined. Here are some examples:

- How to protect life
- When to notify emergency services
- When to notify management
- How to assess the disaster
- How to minimize the disaster's impact

The disaster recovery actions are very important; they should be given a special section in the plan. Here are some disaster recovery actions:

- Activate the backup procedure
- Move to the backup location
- Notify customers
- Reroute traffic to alternative networks
- Maintain systems and facilities security
- Coordinate with management and authorities
- Assess the damage and the resulting liabilities
- Document all actions
- Prepare a statement of the disaster

Another part of the plan should explain how to move to an alternative site, since this needs some extra work. A third part of the plan should explain how to restore services and return the organization to normal. It might contain several duties according to the type of disaster encountered. This topic is discussed in Chapter 9.

12.3.6 Testing of the plan

Chapter 6 explains how to test the plan:

- The purpose of the test
- The recommended approach to testing

- The frequency of testing
- Scheduled and unscheduled tests
- Approved flexibility in testing
- How to select the test team
- How to verify the testing
- How to conduct the testing
- How to analyze the test results
- How to evaluate the test results
- How to modify or adjust the plan

12.3.7   Maintenance activities

Any plan must be dynamic; it must take care of all the changes that happen in the procedures, the new threats that might face the organization or any changes in the planning environment. The plan must be constantly reviewed and updated, otherwise it might become obsolete and ineffective. Here are some of the topics to consider:

- Changes in the business process
- Changes in the hardware or software
- Changes in the applications
- Changes in the technical or managerial staff
- Changes in the equipment and technology
- Changes in the equipment vendors
- Changes in the types of customer
- Changes in the work environment
- Any other changes that affect the plan

12.3.8   Disaster management training

Training is a vital activity that contributes to the success of the disaster management and recovery activities. Training should be done according to a well-formulated plan, as discussed in Chapter 6.

The main item in a training plan is to train all who will be involved in the disaster management and recovery activities such as the team members themselves, the company management, the customers, the equipment (or service) vendors and the new employees.

The frequency of training is important since all those people must get refresher courses to keep them fit for the job if and when a disaster occurs. The tasks assigned to each person should be rehearsed several times until they are well understood and properly implemented

12.3.9 Preliminary report

The preliminary report is prepared by the disaster team leader and presented to management for final approval. It should contain the following items:

- Initial planning steps:
  — Detailed work plan
  — Schedules of interviews with company personnel
  — Objectives and goals of the plan
  — Proposed program to increase disaster awareness between company personnel

- Vulnerability search, analysis and rectification:
  — Security assessment report
  — Potential disasters and their relative importance
  — Legal liabilities
  — Scope and framework of the disaster management plan
  — Preliminary recommendations for recovery strategies

- Disaster impact analysis:
  — Expected impact report (technical and financial)

- Disaster management requirements:
  — Proposed recovery profile for each identified disaster
  — Inventory of available resources
  — Resources and requirements on all timescales

- Development of the disaster management plan:
  — Integrated disaster recovery plan
  — Recovery plan for each anticipated disaster
  — Business resumption plan for each disaster
  — Worst-case scenario for recovery

- Procedure for testing the plan:
  — Testing methods and procedures to implement
  — How testing would satisfy the plan's goals
  — How testing would satisfy the plan's strategies

- Maintenance activities:
  — Schedule of maintenance activities or procedures
  — Monitoring procedures to detect changes
  — Updating the plan to meet new environments

- Disaster management training:
  — Company personnel to be trained
  — Training courses to be given
  — Training methodologies to be followed
  — Criteria for success to be judged
  — Frequency of courses to be delivered

12.3.10 Appendices

Several appendices should accompany the preliminary plan such as:

- Disaster management team members and the tasks assigned to each of them

- Disaster management communications plan

- Cellular telephone numbers of key personnel

- Plan of the buildings:
  — floor layouts
  — escape routes
  — firefighting equipment

- Resource inventories

- Maintenance forms completed and updated

- List of equipment and systems in use

- List of redundant equipment and systems

- Hardware and software license numbers

- Operating instructions of equipment and systems

- Telephone numbers for equipment vendors and service companies

- Network configuration and schematics

- Insurance policies and alternative service agreements

12.3.11 Presentation of the report

The preliminary report is presented to the management and discussed fully (Chapter 6). All modifications approved during the discussion should be incorporated into the final report. The final report is presented to the management for approval.

12.4 IMPLEMENTATION

12.4.1 General guidelines

The disaster management team may face logistical problems such as managing and controlling the flow of information. The best way to insure proper and speedy implementation is by using forms. Here are some suggestions:

- General hazard analysis card
- Detailed hazard analysis card
- Vulnerability search, analysis and rectification form
- Questionnaire for faults causing service interruption
- Criticality of the duration of a disaster
- Criticality of the priority of a disaster
- Criticality index form
- Decision to upgrade facilities and buildings form
- Review of the disaster recovery plan
- Testing of the disaster recovery plan
- Cost-effectiveness of recovery plans for different disasters

12.4.2 Acquiring the resources

Resource acquisition begins as soon as the disaster management plan has been approved. Some resources may already be available at the organization, others must be obtained from outside sources.

12.4.3 Selecting the team

It is seldom that every organizational department is represented in the disaster management team, otherwise it would be too large and too difficult to control. A more efficient way is to select representatives from key areas where disaster probability is high. These people can be easily trained to do the needed tasks, since they have a good knowledge of the relevant processes and an intimate knowledge of the vital details. Another way to start work is to use disaster management consultants. This solution has many advantages:

- They have vast experience in the field, so the project can have a quick start if available time is short.

- They will give the organization several alternative plans and compare the situation with other organizations for whom they have previously prepared plans.

- They are neutral unbiased professionals who will give the organization an objective viewpoint.

It also has disadvantages:

- The cost of consultants is generally high.

- Some vital knowledge of the organization's records might be reviewed by the consultants.

- They might not have the time or desire to train the organization's employees to do the job.

12.4.4 Training the team

The team are trained on the performance of their duties. The tasks given should be clearly defined, realistic and simple to understand. The assignments they have to perform should be accomplished in a reasonable time period — one week is the limit — to ease follow-up and effective control. The result of assignments must be measurable and the criteria of success must be fully understood by the team members before attempting to perform any assignment.

12.4.5 Disaster management drills

It is very important to train the organization's employees on how to behave during a disaster. Disaster management drills should be conducted at least twice per year for disasters that have high priority. These drills should be systematically conducted and properly planned so they do not cause any harm to the organization or the staff. They must be taken very seriously by all concerned. Failure to participate in these drills should be confronted by management at once.

12.5 CASE STUDIES AND LESSONS LEARNT

12.5.1 Hinsdale central office

What happened?

The Hinsdale central office (CO) is an unattended CO which went through a major upgrade to increase its traffic-handling capacity. The contractor who performed the upgrade destroyed the metal cable sheath of some power cables. This was not detected at that time.

On May 8, 1988 a thunderstorm in the area caused several alarms to be sent back to the monitoring station, but the shift technician cleared them as he thought they were the result of the thunder strike's activity. When the number of alarms increased, the technician went to the CO to see what was wrong. When he arrived it was too late because Hinsdale CO was on fire. There was no way he could get to the portable extinguishing equipment or to a telephone to call for help. He couldn't enter the building and all telephone lines were disrupted by the fire. The technician asked a passerby to inform the fire brigade.

When the fire brigade came at last, they searched for the main electricity switch to disconnect the electrical supply. When they switched off the supply, the backup system was energized automatically and extinguishing efforts were hampered by the live equipment. The fire brigade finally switched off the backup power supply and began spraying the equipment with water.

Burnt cables produced intoxicating fumes when they were sprayed with water, and equipment became damaged or unusable. It wasn't only water that damaged equipment, there was also damage from drips of PVC cable sheathing that melted in the fire.

What went wrong?

- Not detecting the destruction of the cable sheath meant that electrical continuity was lost. When the thunderstorm happened, sparks were picked up around the fragmented metal sheath of the cable since it was not properly earthed. This was the main cause of the fire.

- Ignoring or misinterpreting the early warning signs (Chapter 2) led the technician to take the wrong decision. He thought the warning signs were false but in fact they were real. Had the technician taken the matter seriously and checked the situation at once, a major disaster could have been avoided.

- Delay in confronting the disaster reduced the chances of controlling it.

- When the technician arrived, he could not reach the portable fire-fighting equipment (Chapter 9).

- When the technician arrived, he could not reach the emergency telephone (Chapter 9). He should have been equipped with a mobile phone when going to investigate a disaster at a remote unattended CO (Chapter 9).

- The fireman's switch (main power switch) should have been located immediately outside the building and it should have been clearly marked (in red) so that the fire brigade could locate it immediately (Chapter 9).

- The fireman's switch should have been connected to interrupt all power supplies: main, auxiliary and backup. This would have helped the fire brigade, reduced casualties and reduced losses (Chapter 9).

- All the delays allowed the fire to spread for a long time, so the fire brigade had to use water to control it. This was able to stop the fire but it had disastrous effects on the equipment, even items which were not destroyed in the fire. By using carbon dioxide at an earlier stage, many of the losses could have been avoided (Chapter 9).

- Due to the delay in putting out the fire, cables were burnt and equipment was destroyed. Drips of PVC cable sheathing also contributed to the damage of equipment.

What were the real reasons?

- Lack of a disaster management and recovery plan

- Technician not properly trained

- Absence of a disaster communications plan

- Absence of a firefighting plan

What did they learn?

- A disaster management and recovery plan is very important.

- Even without a plan, properly trained employees can minimize the damage.

12.5.2 Roaming interruption on GSM

What happened?

The Telaria telecommunications organization introduced the GSM mobile telephone service in the country and demand was increasing to provide roaming facilities for subscribers when they traveled to neighboring countries. The negotiations were successful and a roaming agreement was established between Telaria's operator and the operators in neighboring countries. Telaria's citizens were very happy to be able to communicate with their home country while visiting neighboring countries. One day Telarian subscribers roaming in a neighboring country were unable to access the visited GSM network. But subscribers who arrived only the day before were able to access the visited network. The same problem was observed by people from the neighboring country on a visit to Telaria. The problem continued for about eight hours, with major losses for the two telecommunications operators in both countries.

What went wrong?

There was an interruption to a signaling link in international signaling system 7 (SS7); this link connected the Telaria GSM network to the neighboring GSM network. The interruption of the SS7 link made it impossible for the roaming subscribers to perform location updates when registering into the neighboring GSM network. The problem was compounded by the unavailability of a backup exchange to provide SS7 signaling.

What were the real reasons?

- An E1 signaling link had a location updates problem, which was due to a hardware problem in the international gateway.

- There was no backup for the signaling network to be inserted in the network when the main SS7 signaling link failed.

What did they learn?

- The cause of the problem was rectified by changing the card in the switch.

- The backup system was initiated by securing the link E1 as a disaster management measure so this problem would not happen again.

12.5.3 Access interruption to GSM HLR

What happened?

The Telaria telecommunications organization introduced the GSM mobile telephone service in the country and demand was increasing at a very good rate. The organization was in the process of negotiating with the manufacturers to increase the capacity of the system when one day some subscribers of the GSM network, who had been on the move within that hour, were unable to access their network. The problem was aggravated because the service interruption occurred during the busy hour. Thousands of subscribers could neither receive nor initiate calls during that period. The telecommunications organization lost large amounts of money.

What went wrong?

Access was interrupted to the network's home location register (HLR). Users on the move during that period could not update their location because an update requires access to the HLR. Thus they couldn't use their phones and they were unable to benefit from the GSM service. Since they were mostly businesspeople, their businesses suffered enormously.

What were the real reasons?

A young service engineer, while loading software updates at the HLR, made an error due to lack of training. The software updates should have been loaded by senior engineers, not a young service engineer. It was not his duty. This indicates a relaxation in the personnel control at the Telaria telecommunications organization.

What did they learn?

- The tasks assigned to each category of engineers should be respected and any violation should be confronted immediately.

- Disaster management measures were taken to insure this problem would not happen again:
  — clearly defined maintenance procedures
  — training courses to raise staff expertise

SUMMARY

This chapter wraps up the subject of disaster management in telecommunications systems. Remember the vulnerability search, analysis and rectification process. Remember the planning process and the preliminary report. After a disaster, ask yourself four questions:

- What happened?

- What went wrong?

- What were the real reasons?

- What have I learned?

REVIEW QUESTIONS

1. What are the main components of the plan preliminary report?

2. How can you identify the vulnerabilities at your workplace?

3. What is the highest disaster risk you face at your workplace? How do you intend to act and why?

4. How can you test your disaster management plan?

5. How can you train the disaster management team?

6. How can you solve the case studies given in this chapter using the forms given in Chapter 11?

7. How can you modify the problem analysis and solution forms given in Chapter 11 to help you solve a case study when the available information is not enough?

Appendices

A

Troubleshooting Transistors

Junction currents and voltages

Measure the base–emitter junction to see if the transistor is on (Figure A.1). If it is, measure the emitter–collector voltage to see if it is small enough. It must be a lot less than the supply voltage. If the base voltage is 10 V, something must be wrong since the collector is not obeying the command sent to the transistor. This indicates a bad transistor, probably an open-base transistor.

Modify the circuit

Try to modify the circuit to enable clarifying the fault. Put a short circuit on the base–emitter junction. If the collector voltage is not equal to the supply voltage, the transistor is leaky. This test must be done in an AC-coupled circuit only. It should not be done in a DC-coupled circuit.

Leakage current

We can use the leakage current as a guide, since leakage current is very sensitive to temperature. A leaky transistor will behave normally when cold but it will start leaking as it gets hotter. Heating and cooling the transistor will reveal whether it is leaky.

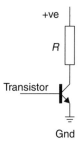

Figure A.1   Junction transistor

DC-coupled circuits

When servicing amplifiers, a leaky transistor can cause distortion. But in DC-coupled circuits the distortion can occur in some stage far ahead of the leaky transistor.

Measure resistance

Check the resistance with an ohmmeter. Use the safe region of the meter. The 1 kΩ range gives very low short-circuit current and an open-circuit voltage. Never use an ohmmeter to check microwave transistors.

Differential amplifiers

Differential amplifiers can be checked by putting the same signal on both the inputs of the amplifier; there should not be any difference between the output terminals (the collectors). If there are several stages in the amplifier, short the inputs of the first stage and verify there is no difference signal in the final stage meter. Then short the second stage and verify. Repeat the process until you find a difference which determines the faulty stage.

Transistor checkers

A transistor checker checks transistors quickly. Use the circuit in Figure A.2 to check each transistor as two separate diodes. Connect the V terminal to

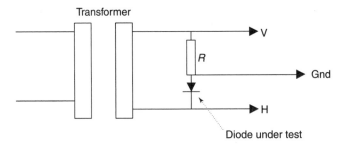

Figure A.2 Transistor checker

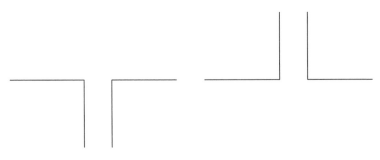

Figure A.3 A good diode gives one of these traces

the vertical plates of the CRO and the H terminal to the horizontal plates. Examine the oscilloscope trace:

- A good diode will give any of the traces in Figure A.3; it depends on the orientation of the plates.

- If the diode is short-circuited, the trace will be a vertical line.

- If the diode is open-circuited, the trace will be a horizontal line.

- Inductance and capacitance may cause the traces to deviate from perfect right angles or perfect straight lines.

B

Troubleshooting Logic Circuits

Logic ICs are divided into two main groups. CMOS ICs use MOSFET transistors to construct logic gates, flip-flops, etc. TTL ICs use bipolar transistors to construct logic components. When troubleshooting CMOS ICs we use a hand strap to connect ourselves to the earth (cold water pipe) through a 500 Ω resistor and also we place the ICs on a metallic sheet connected to earth (the same cold water pipe). TTL ICs work with a supply voltage of 5 V, whereas CMOS ICs work with 3–18 V Avoid using supply voltages that are too high.

B.1 TROUBLESHOOTING LOGIC ICs

1. Take the IC out of circuit. If it is soldered, avoid applying excessive heat to the pins; there are special irons that apply the heat and suck the solder from the pins at the same time. If you want to install it again, or put in a spare IC, use a socket to avoid heating the IC in the future.

2. Check the supply with an ohmmeter to see if you have a short circuit between the positive and the ground. If you have a short circuit, the IC is defective.

3. Check the truth table of the IC. Remember that 0 = GND, 1 = 5 V (TTL) or 3–18 V (CMOS).

A	B	OUT
0	0	0
0	1	0
1	0	0
1	1	1

The IC in our example contains four AND gates. Check each gate alone; take the input and apply either 1's (positive voltage) or 0's (ground voltage). The value of the positive voltage will depend on the kind of IC.

We can use an ohmmeter or an oscilloscope to see whether the gates obey the truth table. If one of them gives different results, the IC is defective. This is a lengthy procedure, but there are computerized systems that can do it in a fraction of a second; however, they are rather expensive. They have libraries containing most of the known ICs. All we have to do is to install the IC in the system, enter its number and wait to see whether or not it's defective.

B.2   FAULTFINDING CIRCUITS WITH MANY ICs

1. Visually inspect the circuit; look for burns, cuts in the tracks of the printed circuit, or a bad contact. If you suspect a tiny cut, measure the continuity with an ohmmeter; if it gives you an open circuit, there is a cut.

2. Try to clean the circuit board, because moisture and sand can produce unwanted resistance in unwanted places. Take the ICs out of their sockets then put them back in; this gets rid of possible bad contacts on the pins. Then try the circuit once more to see if it's going to work.

3. If the circuit is simple, we can apply known inputs and follow the outputs through the different stages in the circuit. Any stage that gives an unexpected output is defective.

4. If the circuit is really complex, begin by checking the large ICs. This is because large ICs can contain thousands of transistors, so they are more susceptible to failure than discrete components. After checking the large ICs, check the smaller ones, and so on.

5. Large ICs, e.g. microprocessors, are difficult to check without a computerized system. If you don't have a computerized system, try to replace the IC with a new one. A new IC usually costs less than paying you to test the old one.

6. For analog ICs it is better to use a computerized system. If you don't have a computerized system, try these procedures. Check for short circuit on the supply. Measure the amplification of an op-amp. Apply AC and measure the op-amp's output frequency and slew rate.

C

Troubleshooting FETs and MOSFETs

Cleanliness

The junction FET acts as a variable resistance; by varying the bias we control the channel current. The MOSFET gate looks like a capacitor, having higher input impedance than the JFET gate, which looks like a diode (Figures C.1 and C.2).

Before suspecting a FET, clean the board around the legs of the FET to avoid leakage traces. Before changing a FET, clean both sides of the board.

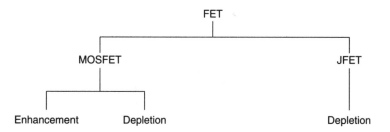

Figure C.1 Types of FET

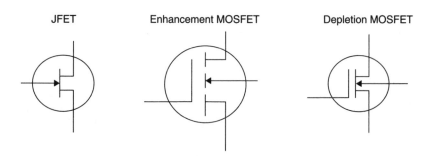

Figure C.2 FET symbols

Clean it with a stiff brush then spray the FET and the area around it with Freon degreaser. Test the circuit and if all is okay, spray with a lacquer sealer.

In-circuit testing

Probing the FET source (not the gate) will show whether a signal is present. Probing the drain will reveal any gain (amplification). The voltage gain in FETs is 5–25 (low), whereas in transistors it is more than 200 (high).

Differential amplifiers

Troubleshooting FETs is easy if there is a means to adjust the zero. Put the voltmeter probes between the two drains (on the output) and vary the zero adjust; if all is well, the meter will swing both sides as you vary the zero adjust.

Leakage current

Leakage current can upset FET circuit operation; it occurs mainly with JFETs. Leakage current can lower the input impedance. The leakage current is $E/R_g < 1000$ pA for a good JFET.

Gate control for channel current

To check the gate, measure the source–drain DC voltage then short the gate with the source; in a good JFET this will make it conduct more and will lower the source–drain voltage.

MOSFET handling

Handle MOSFETs with care; never allow the gate to float, else it might pick up static charges that puncture its thin dielectric film.

Don't insert the MOSFET in a block of Styrofoam; this can generate enough static electricity to ruin it. Each MOSFET comes with a wire connecting its leads together; don't remove this wire until all leads are connected in your circuit. This is not so critical on a JFET.

When soldering a MOSFET into the circuit, use a grounded soldering iron, or connect the iron to a good ground, not a cold water pipe. Don't use the soldering gun (transformer); it might induce a voltage spike that ruins the MOSFET.

Out-of-circuit tests

Out-of-circuit tests can be performed on JFETs with an ohmmeter. Put diodes between gate–source junctions (high resistance in one direction and low resistance in the other) and between gate–drain junctions (the drain–source channel resistance has a low value of 100 to 10 000 Ω). These tests tell us if we have an open circuit or a short circuit. Never use an ohmmeter to test MOSFETs since this would require you to remove the protective wire that connects the leads together.

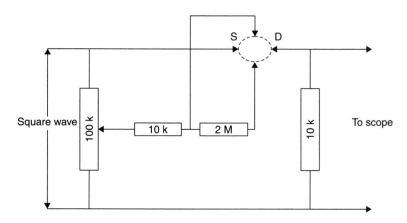

Figure C.3 FET checker

FET checker

The circuit in Figure C.3 can be used to check all types of FET out of circuit.

Output level

Using an enhanced n-channel FET, for example, look at the oscilloscope trace for a pulse train of square waves in the output. Change the gate voltage; if the output level varies, you have a good FET.

D

How to Present Data for QoS Figures

Global quality-of-service index

Item	1	2	3	4	5	6	7	8	9	10	11	12
Objective												
Realized												

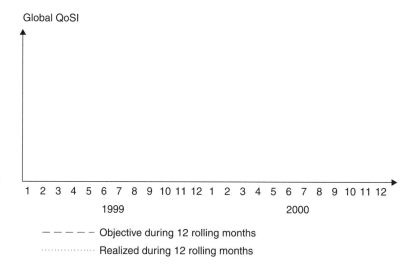

Monthly fault rate (MFR)

Month	Monthly result (MFR)	Average during 12 rolling months	
		Objective	Realized
1			
2			
3			
4			
5			
6			
7			
8			
9			
10			
11			
12			

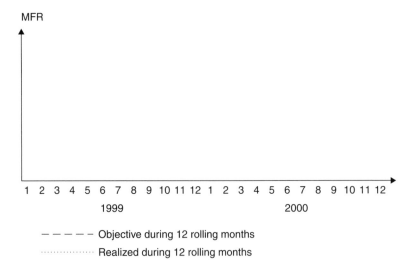

MFR

1 2 3 4 5 6 7 8 9 10 11 12 1 2 3 4 5 6 7 8 9 10 11 12
        1999   2000

– – – – – Objective during 12 rolling months
·················· Realized during 12 rolling months

E

Paralleling of Broadcasting Transmitters

Transmitting stations connected in parallel provide more power, and if one transmitter is interrupted for any reason, the program can still be broadcast using the other transmitter. Figure E.1 shows that the paralleling unit has unique characteristics, since the inductive and capacitive reactances at resonance (at the operating frequency of the transmitter) must each equal the characteristic impedance of the feeder line, which must also be equal to the resistance of the dummy load:

$$R = 1/2\pi f C = 2\pi f L$$

where f is the transmission frequency. In this case, and with the conditions set forth, there will be no current flowing in the dummy load. The circuit can be simplified to look like Figure E.2. Here the impedance of each transmitter is given by

$$Z = jR + \frac{2R(-j2R)}{2R - j2R} = jR - \frac{j2R}{1 - j}$$

$$= jR\frac{(1 - j) - 2}{1 - j} = R\frac{(j + 1) - 2j}{1 - j} = R$$

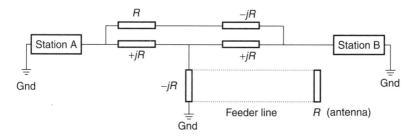

Figure E.1 Paralleling unit: both transmitters working

The impedance will be resistive, so there will be full transmission of power without any reflections. If the power radiated by each transmitter is W, the total radiated power will be $2W$.

If one transmitter is not radiating when both transmitters are connected to the paralleling unit, the circuit can be represented by Figure E.3. This

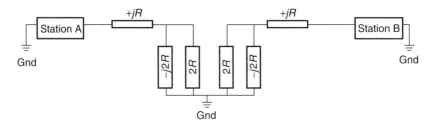

Figure E.2 Simplified circuit: both transmitters working

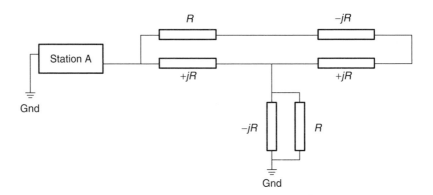

Figure E.3 Paralleling unit: one transmitter down

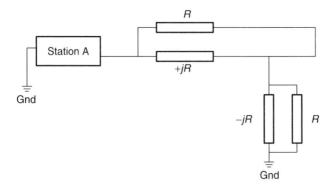

Figure E.4 Simplified circuit: one transmitter down

can be simplified to Figure E.4. Here the impedance is given by

$$Z = \frac{jR^2}{R(1+j)} - \frac{jR^2}{R(1-j)} = jR\frac{(1-j)-(1+j)}{2}$$

$$= jR\frac{-2j}{2} = R$$

The impedance is also resistive, so there will be no reflections. But half the power will be lost in the dummy load, thus the radiated power will be W instead of $2W$. The dummy load cannot dissipate all that power unless it is given proper cooling.

F

Financial Turnaround at a TV Tube Maker

F.1 INTRODUCTION

Manufacturing TV picture tubes is a very delicate business; it needs long training and involves many skills: glass technology, electronics, chemical modification and heat treatment.

When a developing country planned to introduce tube manufacturing, it was thought that buying a manual factory would be cheaper than a semiautomatic or fully automatic plant. It was also considered that a manual factory would serve as a training aid for the engineers and technicians, so they could get a feel for this new industry before the government bought a semiautomatic plant.

To reduce the cost further, the government decided to buy a thirdhand manual factory built in the United Kingdom and sold to Australia then onto Greece. Having worked in both countries for several years, the factory was in bad shape by the time it was delivered.

The factory was installed and operated with the help of expatriates from the manufacturing company, but it never attained its full capacity. After the expatriates left, the situation became worse since some of the best workers found better positions in the glass-manufacturing industry. The production tumbled to 2000 tubes per year instead of the design figure of 150 000 tubes per year. The production cost per tube was £16 (£4 overheads per tube) and the market selling price was £14 per tube.

F.2 STATEMENT OF THE PROBLEM

When I was called in to manage this factory, I began by analyzing the production process.

F.2.1   Process description

- The activities and work steps performed in the TV picture tube manufacturing process.

- The work methods and technical aids used when performing these activities.

- The mutual interdependence of the different activities and work steps.

- The location, with respect to the activities and work steps, of the decision-making points in the process.

- For each decision-making point, identify:
  — the decision criteria
  — the viability of the decision process
  — the decision makers
  — the consequences of each major decision

F.2.2   Production flowchart

Basic production processes

Figure F.1 illustrates the three production processes for producing the TV picture tube from bare glass. Notice there are only three quality checkpoints.

Factory layout

Figure F.2 illustrates the factory layout at the beginning of the mission, and Figure F.3 after the modifications.

Detailed process chart

Figure F.4 illustrates the detailed manufacturing process.

Detailed process functions

- *Washing*: the glass bulb is washed from inside by specially treated water.

- *Hydrofluoric acid*: the bulb face is etched from inside by hydrofluoric acid.

- *Phosphorus precipitation*: phosphorous material is precipitated on the inner face of the bulb.

- *Drying*: the bulb is dried.

- *Lacquering*: a lacquer resin is spread over the inner face of the bulb to help fix the phosphorous material.

- *Aluminizing*: a film of aluminum is sprayed over the lacquered area to help increase the light intensity.

- *Graphite addition*: a conducting material is added in certain areas to help make good electrical connections.

- *Baking*: the bulb enters a large moving-mat baking furnace to help remove all stresses in the glass.

- *Gun washing*: the electron gun is washed in alcohol to insure it is clean.

- *Gun sealing*: the gun is sealed to the neck using a gas flame.

- *Evacuation, activation, sealing*: air is evacuated from the tube using a diffusion vacuum pump inside a furnace. The electron gun in the tube is then evacuated of all residual gases using the vacuum pump under high temperature. The electron gun vent is then sealed using a gas flame.

- *Electronic process*: the tube is activated to get the rated currents then getter-fired and spot-knocked to insure maximum evacuation.

F.2.3 Actual efficiency versus design efficiency

Item	Manufacturer's figures	Achieved figures
Production capacity (tubes/shift)	150	100
Final production (tubes/shift)	100	20
Production efficiency (%)	68	20
Chemical section efficiency (%)	90	70
Thermal section efficiency (%)	80	60
Electronic section efficiency (%)	95	50

F.3 MAIN CAUSES OF THE PROBLEM

F.3.1 Product flow problems

The technological procedures did not run in a smooth sequence (Figure F.2). The production flow and the equipment layout, were therefore modified in the chemical section and the electronic section. This reduced the time to transport the semiprocessed product within the factory and also minimized the defects attributed to bad handling between the different sections (Figure F.3).

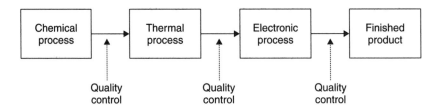

Figure F.1 The original process had only three quality checkpoints

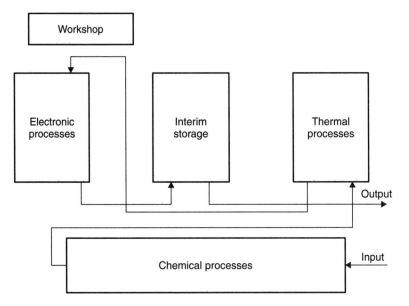

Figure F.2 Factory layout: before I took over

F.3.2 Few quality checkpoints

There were only three quality checkpoints, one at the end of each section. This caused a lot of wasted time and production supply materials, because of unnecessary rework due to defects within the subprocesses performed in each section. A new system was introduced to check the quality after each subprocess and reduce the rejects.

F.3.3 Many inoperative items

- Lack of spares or supplies (oil, chemicals, nickel chrome, etc.).
- Bad handling of semiprocessed product by inexperienced personnel.
- Lack of proper maintenance procedures and experienced maintenance technicians.

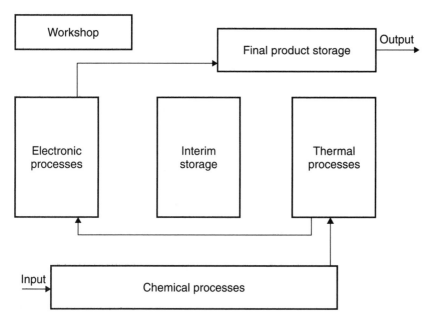

Figure F.3 Factory layout: after I modified it

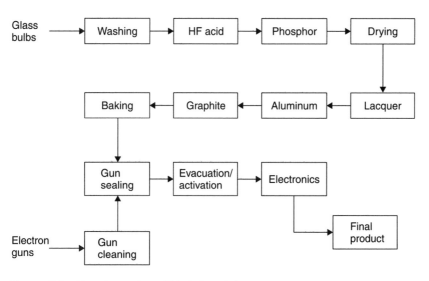

Figure F.4 How to make a TV picture tube

F.3.4   Things that were lacking

- Lack of some vital production supplies such as lacquer (used for inner coating of the screen).

- Lack of production specifications and documentation, such as the process sheet, instrument catalogs and material specs.

- Lack of experienced production personnel.

F.3.5 Very high rework and rejects

- Serious production faults.

- Serious defects in chemical purity.

- Improper treatment of the electron guns.

- Improper thermal treatment of the TV glass.

- Improper sequence of Aquadag (graphite).

- Inefficient and inadequate production flow.

F.3.6 Lack of proper plant conditions

- Dust and rainwater falling on the processed product from the plant roof (a waterproof sheet was used to cover the roof as a quick measure).

- Chemically polluted air within the plant (the ventilation fans had locally manufactured filters).

- High humidity affected the chemical section only. It was solved by air-conditioning the section rooms to insure a stable temperature and humidity, crucial for product quality.

- Improper handling of the bulbs and semifinished tubes left them with scratched faces. The problem was solved by a local modification to the carrying trolley.

- Proper storage for the finished product was created after rerouting the production line.

F.3.7 Serious personnel problems

- The section heads and senior technicians didn't know the process well, yet they were supposed to instruct the operators.

- No one was able to solve production faults that needed some research, so the reject percentage remained high.

- Due to reduced production, the capital cost per tube became excessive, so the tube's selling price increased and became uncompetitive, making the factory lose money.

- The factory personnel became very demoralized because they did not get any bonuses or incentives.

F.4 ACTION PLAN

F.4.1 Setting goals

The short-term goals were to increase production efficiency by reducing rejects at each decision-making point, through efficient and strict quality checks at each subprocess. The long-term goals were to achieve design production capacity at the proper efficiency; this would reduce the overhead costs per tube and make the selling price competitive.

F.4.2 The work plan

The priorities were defined according to the goals that had been set. For the short term it was found economical to shut down the factory for three weeks to perform these four operations:

- All production personnel to attend a crash training program on how to better perform their specific duties. The training was given by senior technicians, themselves trained by the plant's UK manufacturer.

- All maintenance staff to bring into operation the maximum number of machines and equipment by taking vital spare parts from inoperative equipment and listing the major component purchases needed to repair the rest.

- All production manuals to be retrieved from operational staff. They were made available to all personnel by putting them in the factory's library and a copy in each production section. Missing manuals were ordered from the manufacturers.

- All needed supplies to be secured from local suppliers whenever possible. This insured continuity of supply and circumvented problems with foreign currency.

F.5 PROBLEMS AND THEIR SOLUTIONS

F.5.1 Operational problems

Problems that were easy to solve

Repair of inoperative machines

It was found that 80% of the inoperative diffusion vacuum pumps lacked the heating element, an imported item, but with the help of the indigenous clay manufacturers, an equivalent heater was made available. It worked with 75% efficiency, but it cost the factory only 10% of the imported heater's cost and therefore kept down production expenses.

A new quality control system

A new system of quality control was introduced to check the product after each operation, to help reduce the loss, and to identify more precisely the points that generated most rejects. This enabled the production engineer to concentrate on the most vital parts of the process.

Factory cleanliness

Dust was a big problem. A very strict program was introduced to keep the floors and all machines clean.

Problems that needed some research

Phosphorous material

It was found that the radiated light from the tube was not uniform, and the tube was 20–30% less efficient than an equivalent imported tube. Since trials at the factory failed to achieve any result, it was decided that one of the company's chemists would research this problem. They found the correct percentages of the different elements in the phosphorous material and also the required purity. When the results of the research were implemented, the manufactured tubes produced the required uniform light output.

Cathode poisoning

It was noticed that the cathode current in the finished tube was about 25% lower than in the equivalent imported tube. Analysis at the factory's lab indicated that the cathode material was not homogeneous, but further investigation revealed that it became contaminated during the activation process due to the inefficiency of some diffusion pumps. Research was eventually transferred to the local university and the cathode current became normal after implementing these three findings:

- The activation procedure was inappropriate, so a new procedure was proposed.

- The cathode and gettering materials had low purity. Contact with the manufacturer helped to resolve the problem, and subsequent deliveries were up to spec.

- The getter firing was inefficient due to low high-frequency (HF) output power radiated from the HF coil at the factory. This was solved by using a more powerful HF generating unit.

Glass cracks

There was a high reject rate (20–25%) at the output of the thermal section, due to glass cracks at the cone/face area, and also at the neck/gun seal area. The factory's efforts to reduce this high rate did not produce good results, so an engineer from the thermal section conducted some research at the Glass Research Institute. Here are the results:

- The rate of change of temperature inside the long baking furnace was the reason for the cone/face cracks; a new rate was proposed.

- The method of sealing the gun to the neck was the reason for the cracks. A new clay radiator was designed by the engineer. Its parabolic shape helped to produce a more uniform flame distribution around the neck. It gave good test results and helped to reduce rejects to only 2–5%.

Production bottlenecks

There was a major bottleneck between the end of the thermal process and the start of the electronic process, since the output from evacuation and sealing was very low compared to the capacity of the electronic processing unit. The evacuation oven was designed to hold one 27 in (686 mm) tube. The tubes in production at that time were only 16 in (406 mm), so the trolley holding the tube was redesigned. Two new trolleys could fit in the space of one old trolley, doubling the capacity of the oven and reducing its heating costs. This brought down the overall cost of tube production.

F.5.2 Environmental problems

Dust prevention

Dust was a main cause of low efficiency in the chemical section and also in gun preparation and sealing. Dust prevention was achieved in three ways:

- Covering the factory roof (which was made of corrugated metal) with bitumen-coated sheets.

- Installing a few water sprinklers on the factory ceiling to help eliminate residual dust particles in the air.

- Using air blowers to blow clean and filtered air into the factory, making the pressure inside a little higher than outside; this helped to prevent dust from entering.

Temperature and humidity

The change in temperature in the phosphorus process (chemical section) disturbed the homogeneity of the screen layer. This led to reduced light emission from the TV tube. This problem was solved by fitting the phosphorus room with a simple thermostat, designed and built in the factory's workshop.

F.5.3   Administrative and financial problems

Many problems were faced at the outset, yet the main ones were solved when the production efficiency increased and the factory became profitable. The factory personnel were given better pay and more incentives for their higher efficiency. The capital cost per tube went down because more tubes were being produced.

F.6   ACHIEVEMENTS

After five years of very hard work, the factory achieved a yearly production of 100 000 tubes on 3 shifts per day, 5 days per week, 50 weeks per year. Here are the figures.

Item	Manufacturer's figures	Achieved figures
Production capacity (tubes/shift)	150	150
Final production (tubes/shift)	100	133
Production efficiency (%)	68	88
Chemical section efficiency (%)	90	96
Thermal section efficiency (%)	80	95
Electronic section efficiency (%)	95	97

F.7   LESSONS LEARNT

- Solving production problems using a scientific (technical and financial) approach to develop and use suitable applied technologies, this is the best way to achieve results and save valuable production time.

- Cooperation between industry and universities or research institutes is vital in solving problems that need a thorough in-depth investigation. Sophisticated test and measuring equipment is often essential.

- Companies and universities can profit from collaboration.

- Administrative and financial problems should not divert the plant manager from the main goal—achieving higher production efficiency.

- On-the-job training is essential for higher production efficiency.

Bibliography

BOOKS

- J. Adair (1987) *Not Bosses But Leaders*, Talbot Adair Press, Newlands, Surrey.

- R. J. Bates (1992) *Disaster Recovery Planning*, McGraw Hill, Watsonville, CA.

- Bell Telephone Laboratories (1978) *Engineering and Operation in The Bell System* (Second Printing) Bell Telephone Laboratories Inc.

- F. C. Cuny (1983) *Disasters and Development*, Oxford University Press, Oxford.

- S. Fink (1986) *Crisis Management*, American Management Association, New York.

- C. H. Gibson (1992) *Financial Statement Analysis Using Financial Accounting Information* (Sixth ed.), South Western Publishing, Ohio.

- R. Hays and R. Watts (1986) *Corporate Revolution, Heinemann*, London.

- IEE (1992) *Wiring Regulations BS 7671* (Sixteenth ed.), IEE London.

- ITU (1979) *Training Development Guidelines*, ITU, Geneva.

- ITU (1999) *Trends in Telecommunication Reform 1999*, ITU, Geneva.

- A. Jay (1967) *Management and Machiavelli*, Penguin Books Ltd, London.

- L. Kappelman (ed.) (1997) *Year 2000 Problem Strategies and Solutions from the Fortune 2000*, International Thompson Computer Press, Boston.

- J. Keogh (1997) *Solving the Year 2000 Problem*, AP Professional, Boston.

- O. P. Kharbanda and E. A. Stallworthy (1987) *Company Rescue*, Heinemann, London.

- J. R. M. Kunz (1987) *The American Medical Association Family Medical Guide*, Random House, New York.

- A. M. Levitt (1997) *Disaster Planning and Recovery: A Guide for Facility Professionals*, John Wiley & Sons, New York.

- R. E. Linneman (1980) *Shirt-Sleeve Approach to Long-Range Planning*, Prentice-Hall, Inc., Englewood Cliffs, NY.

- J. Maxwell Adams (1994) *Electrical Safety. Power Series 19.* IEE, London.

- G. Meyers (1986) *Managing Crisis*, Unwin Hyman, London.

- C. Northcote and N. Rowe (1979) *Communicate*, Pan Books Ltd, London.

- C. Parker and T. Case (1993) *Management Information Systems* (Second ed.) McGraw Hill, Watsonville, CA.

- B. Ragland (1997) *The Year 2000 Problem Solver*, McGraw Hill, New York.

- M. Register (1989) *Crisis Management*, Hutchinson Business, London.

- S. Tabbane (2000) *Handbook of Mobile Radio Networks*, Artech House Inc., Norwood, MA.

- G. R. Terry (1972) *Principles of Management* (Sixth ed.), Richard D. Irwin, Homewood, IL.

- R. H. Thouless (1974) *Straight and Crooked Thinking*, Pan Books Ltd, London.

- J. W. Toigo (1995) *Disaster Recovery Planning: For Computers and Communication Resources*, John Wiley & Sons, New York.

- J. Van Duuren, P. Kastelein and F. C. Schoute (1992) *Telecommunication Networks and Services*, Addison Wesley, Harlow.

- L. A. Worbel (1990) *Disaster Recovery Planning for Telecommunications*, Artech House, Norwood, MA.

CONFERENCE PAPERS AND MAGAZINE ARTICLES

- G. El Mahdy (1996) Manufacturing process diagnosis and development—a case study. Second International Conference on Manufacturing Processes, Systems and Operations Management in Less Industrialized Regions, Bulawayo, Zimbabwe.

- A. Schwartzelmuller (1995) Effect of underground water on frequency of lightning strikes. University of Zimbabwe, Harare.

- A. Schwartzelmuller (1994) Study of lightning strikes to rural huts in Zimbabwe. University of Zimbabwe, Harare.

- A. Schwartzelmuller (1997) Prevention of lightning injuries. University of Zimbabwe, Harare.

- M. J. Ranum and M. Curtin (1998) Internet firewalls: frequently asked questions.

- International Review of Criminal Policy. United Nations manual on the prevention and control of computer-related crime.

- G. Thurston, P. Russo and L. Lang (2000) Visions for a new public network. *Telecommunications*, Jan 2000.

- A. Harrison (2000) Corporate security begins at home.

- Challenge to the network. Internet for Development, ITU, Geneva (1999).

- Trends in telecommunication reform 1999. Convergence and Regulation, ITU, Geneva (1999).

- C. O'Hara (1999) USPS can't promise mail delivery won't encounter bugs. *Federal Computer Week*, Feb 23, 1999.

- P. Thibodeau (1999) Congress okays Y2K bill. Computerworld Online News, Feb 7, 1999.

- Anon (2000) Senators want to double Hacker penalties. Yahoo News, Feb 18, 2000.

- Anon (2000) Conduct a Y2K post-mortem. IT Agenda 2000 online report.

- American Red Cross (1993) Emergency preparedness checklist.

- American Red Cross (1993) Your family disaster supplies kit.

- Federal Agency Management Agency (1997) Floods and flash floods factsheet. FAMA, March 18, 1997.

- Regional Workshop on the Year 2000 Problem and Telecommunications. ITU, Cairo, Dec 1998.

- Computers and Communications 2000. *Newsweek* supplement, Feb 28, 2000.

- Computer Security Institute (2000) Computer crime and security survey.

- D. Lehman (2000) Able cuts ground Northwest flights. *Computerworld*, March 22, 2000.

- A. Harrison (2000) Cost of cyberattacks rises. *Computerworld*, March 22, 2000.

- L. Rosencrance (2000) Software errors release TWA customers' e-mail addresses. *Computerworld*, March 22, 2000.
- A. Radding (1999) Disaster. *Computerworld*, Jan 1, 1999.
- Anon (1999) Tips for prepping data center disaster recovery teams. *Computerworld*, May 1, 1999.
- Anon (1999) Storm tests IT backup plans. *Computerworld*, Feb 9, 1999.

VIDEO TRAINING PROGRAMS

- Motivating people in today's workplace. Career Track.
- How to delegate work and ensure it is done well. Career Track.
- Dealing with conflict and confrontations. Career Track.
- Accidents don't happen in the office. Training Media Group.
- Safety signs. CITB Construction, Bricham Newton Training Centre.
- Finance for the non-financial manager. TV Choice Productions.
- Coaching for results. BBC Business.
- Nine traits for highly successful work teams. Career Track.
- Practical budgeting skills for managers. Career Track.
- The ten sins of communication. Video Communications Pty Ltd.
- Negotiation. Training Communications Ltd.
- Team building. Career Track.

Contact addresses

- Career Track Publications, 3085 Center Green Drive, Boulder Co 80301, USA.
- Training Media Group, 3a Station Parade, Ealing Road, Northolt, Middlesex UB5 5HR.
- CITB Constructions, Bricham Newton Training Centre, Bricham Newton, Kings Lynn, Norfolk PE31 6HR.
- TV Choice Productions, 22 Charing Cross Road, London WC2H 0HR.
- BBC Business, BBC Enterprises, Woodlands, 80 Wood Lane, London W12 0TT.
- Video Communications Pty, and Training Communications Ltd, Brooklands House, 29 Hygate, Werrington, Peterborough, Cambridgeshire PE4 7ZP.

COMPANY LITERATURE

- FETEX-150 digital switching system. System description for the FXE0150-6106-21, Fujitsu, Japan.

Index